Features of the Near Side Moon

John Moore

W0007380

Image Credits & Acknowledgements

LRO and LROC

Except where otherwise noted, all featured images used in this book are credited to the organisations, institutes and universities involved with the Lunar Reconnaissance Orbiter Cameras (LROCs) and the Lunar Reconnaissance Orbiter (LRO).

For more on LRO and LROC see:
NASA Goddard Space Flight Center: http://lro.gsfc.nasa.gov/
Lunar Reconnaissance Orbiter Camera: http://lroc.sese.asu.edu/index.html

International Astronomical Union (IAU)

Nomenclature, latitude and longitude coordinates for features given in this book are from the official International Astronomical Union (IAU) Working Group for Planetary System Nomenclature (WGPSN).

For more on IAU see:
Gazetteer of Planetary Nomenclature (IAU): http://planetarynames.wr.usgs.gov/

Lunar Terminator Visualization Tool (LTVT)

LTVT was used to project the aerial, gridded images provided in this book.

For more on LTVT see:
http://ltvt.wikispaces.com/LTVT

Features of the Near Side Moon
ISBN-13: 978-1502944955
ISBN-10: 1502944952

Contents

Contents

Features and Book Layout

The most obvious features on the Moon are its craters and huge basins. Easily observable through any optical device, let alone naked eye, too, there are, however, other features that really don't get that much of a 'look-in'. Why? Simply, because the majority of them are not generally well-known.

Of course, there are the odd popular ones and twos: for example, say, the fault-like feature of Rupes Recta, the mountain ranges around the Imbrium Basin, or the promontory-like feature of Laplace (fig 1).

Fig 1. Left is Rupes Recta, middle Montes Carpatus and right is Promontorium Laplace.

However, where would features like the Dorsa Burnet ridge, the Catena Yuri crater chain, or the rille of Rima Cleopatra (fig 2) be on the Moon?

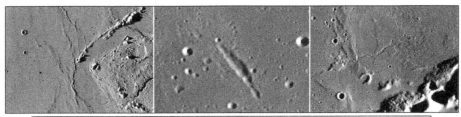

Fig 2. Left is Dorsa Burnet, middle Catena Yuri and right shows Rima Cleopatra.

And those above are just a few in the official list of features as laid down by the International Astronomical Union (IAU) - the organisation responsible for naming features on planetary bodies within our Solar System. But what about others, like those in the unofficial list of features - from domes to ghost craters to dark deposits (fig 3) - that will prove even more problematic when trying to discover them.

Fig 3. Left shows domes in Mon Rümker, middle that of ghost crater Wolf T and right the dark deposits south of Montes Apenninus (some of the most impressive on the Moon).

Features and Book Layout

Features of the Near Side Moon has therefore been produced to simplify and address all of the above. Divided into two sections: the first section involves all the official features (with official designations like latitude, longitude and size) found in the IAU list; while the second section provides those unofficial features that today remain without deserved status. Each feature, therefore, in both sections is looked at from a 'zooming-in' process (fig 4); where, firstly, its

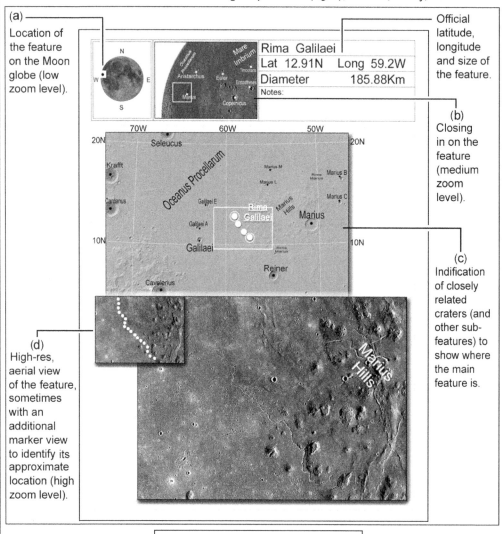

(a) — Location of the feature on the Moon globe (low zoom level).

Official latitude, longitude and size of the feature.

(b) Closing in on the feature (medium zoom level).

(c) Indication of closely related craters (and other sub-features) to show where the main feature is.

(d) High-res, aerial view of the feature, sometimes with an additional marker view to identify its approximate location (high zoom level).

Rima Galilaei
Lat 12.91N   Long 59.2W
Diameter   185.88Km
Notes:

Fig 4. Image layout for locating a feature.

location is shown generally on a global image of the Moon (fig 4a), followed by an additional series of images (fig 4b - d) that brings the feature into view at a more detailed level.

Features and Book Layout

The main objective in producing this book, therefore, is to have an easy-to-find image of the main feature being sought after. The official list of features are listed similar to how they appear in the IAU list; each following along the alphabetic convention. Note, when looking up the main feature, see also other features that lie nearby - supplied to, perhaps, give a bigger-picture view of the possible geological sequence of events that occurred in the main featured area. Initially, the unofficial list of features presented weren't going to be part of this book, so those provided are there to serve as an introduction.

Nomenclature of lunar features
Naming of features for the Moon is carried out by a Working Group at the IAU whose job it is to review suggestions, and then approve/disapprove them. Most names are usually proposed by professional investigators mapping the Moon and its geological features, however, there is also a situation where submissions from the public are considered, too.

The first systematic naming of features happened in 1935 and put into a report known as 'Named Lunar Formations' - compiled by Mary Blagg and Karl Müller. The list back then attributed names to, usually, deceased scientists and philosophers, followed with artists, writers and mythical characters. As more and more knowledge came in about the Moon through better instruments, and later from spacecraft images, the list grew, and so changes to things like heights and sizes of specific features had to be updated.

One or two features, as well as conventions of designations, have been dropped in the current IAU list, however, as there are so many features that need names today the list is sure to grow and change in the near future. Below (fig 5), is the current list of official IAU features shown in the book (with unofficial, too).

OFFICIAL LIST		UNOFFICIAL LIST	
Feature	Amount	Feature	Amount
Catenae (crater chains)	11	Concentric Craters	11
Dorsa (ridge)	18	Dark Halo Craters	11
Dorsum (ridges)	21	Domes	11
Lacus (lakes)	17	Floor Fractured Craters	11
Maria (seas)	20	Ghost Craters	11
Mons (mountain)	25	Miscellaneous	8
Montes (mountains)	19	Rayed Craters	12
Palus (marsh)	3	Simultaneous Impact Craters	5
Promontorium (promontory)	9	Swirls	5
Rima (rille)	52		
Rimae (rilles)	58		
Rupes (fault)	8		
Sinus (bay)	9		
Vallis (valley)	12		

Fig 5. Official and Unofficial list of features

Features and Book Layout

Note: Defining the exact lengths or sizes for some features, for example, where does the edge of a Mare end, or what is the exact end-to-end point of, say, a Catena, isn't always possible at the scale of images produced in this book, so some leeway should be allowed for those presented.

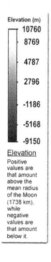

Elevation (m)

10760

8769

4787

2796

-1186

-5168

-9150

Elevation
Positive values are that amount above the mean radius of the Moon (1738 km), while negative values are that amount below it.

To Neil
(the 'bud')

Catenae

Catenae

Catenae is the plural form of the word Catena, which in Latin means 'chain'. When used in a lunar context, the chain refers to a 'line of craters' produced as a result in breakup of an object whose fragmented pieces struck the moon's surface - usually in a linear formation. Most of the crater chains featured here may have formed by the ejecta from larger craters in their vicinity (or, in some cases, may be exterior in origin like an incoming comet), however, others may be structurally related - like the collapse of the surface (small, closely-space pits?) along weaker (usually linear) zones.

The chains' orientation can sometimes help in determining the original source of the crater that may have formed them - if viewed from a *radial* and *secondary impact* perspective, however, for those Catena that may have formed circumferentially outwards, determining their original source will be that much harder. In such instances, two aids may help: firstly, some craters within the chain may have longer axes that 'point' approximately towards the direction of the possible source; secondly, some craters may show a herringbone, v-shaped appendage whose apex again 'points' back to the primary source.

No.	Feature	No.	Feature	No.	Feature
1	Abulfeda	5	Krafft	9	Taruntius
2	Brigitte	6	Littrow	10	Timocharis
3	Davy	7	Pierre	11	Yuri
4	Humboldt	8	Sylvester		

Note: The above list of Catena represents those given in the International Astronomical Union list, but there are plenty more on the Nearside's face that may in the future be given official designations.

5

Catenae

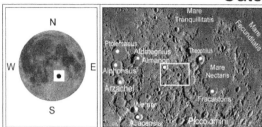

Catena Abulfeda	
Lat 16.59S	Long 16.7E
Diameter	209.97Km
Notes:	

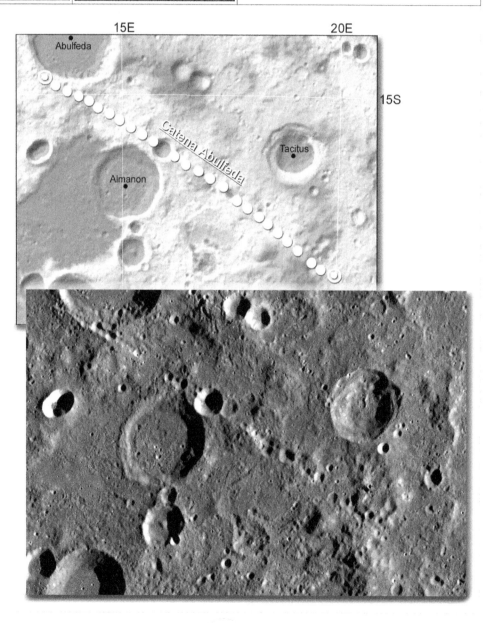

Catenae

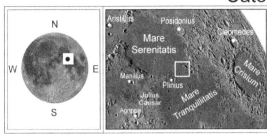

Catena Brigitte	
Lat 18.5N	Long 27.49E
Diameter	7.65Km
Notes:	

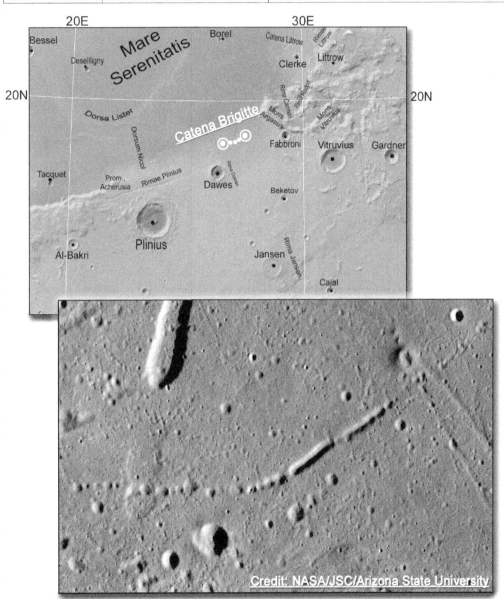

Credit: NASA/JSC/Arizona State University

Catenae

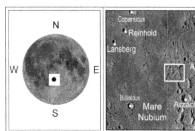

Catena Davy	
Lat 10.98S	Long 6.27W
Diameter	52.34Km
Notes:	

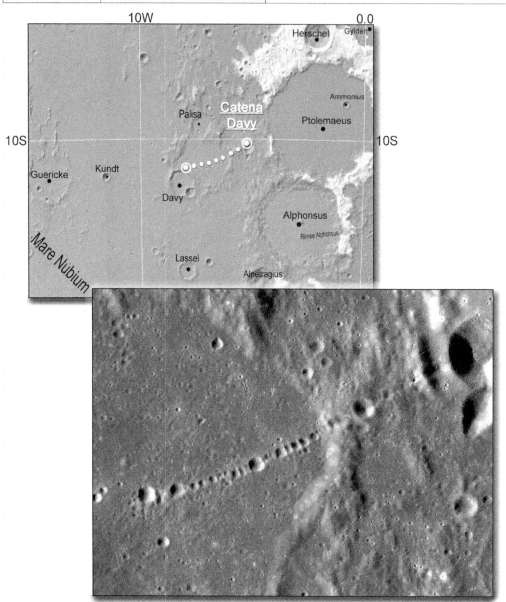

Catenae

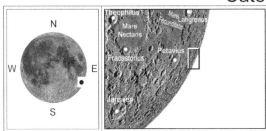

Catena Humboldt	
Lat 21.98S	Long 84.7E
Diameter	162.29Km
Notes:	

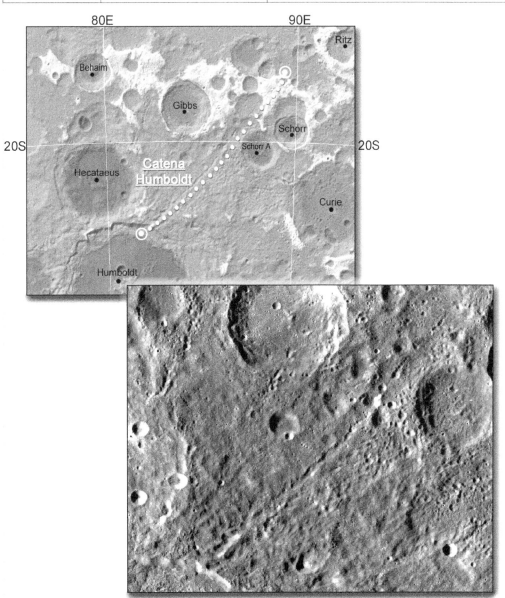

Catenae

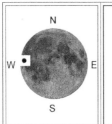 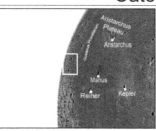

Catena Krafft	
Lat 14.91N	Long 72.25W
Diameter	55.11Km
Notes:	

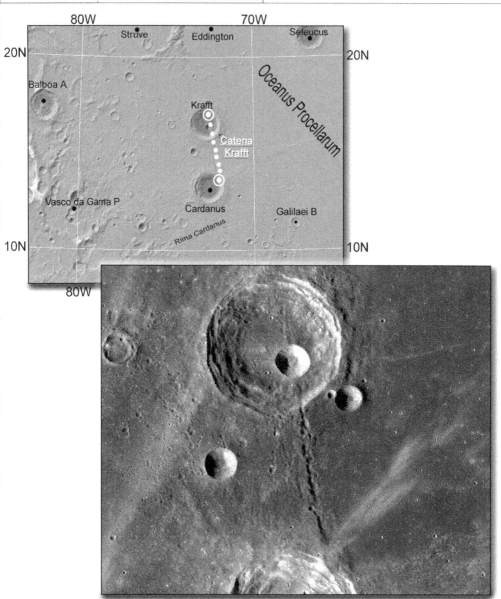

Catenae

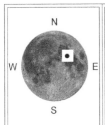

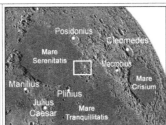

Catena Littrow	
Lat 22.23N	Long 29.61E
Diameter	10.3Km
Notes:	

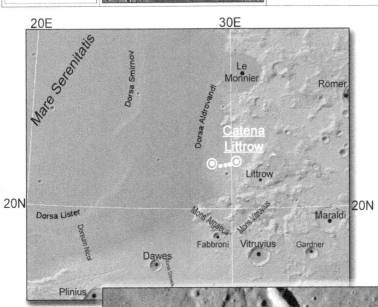

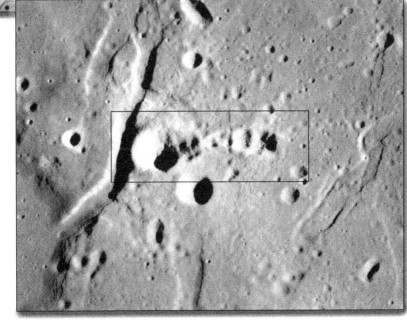

Catenae

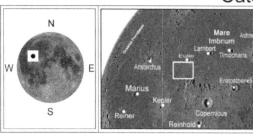

Catena Pierre		
Lat 19.76N	Long	31.86W
Diameter		9.44Km
Notes:		

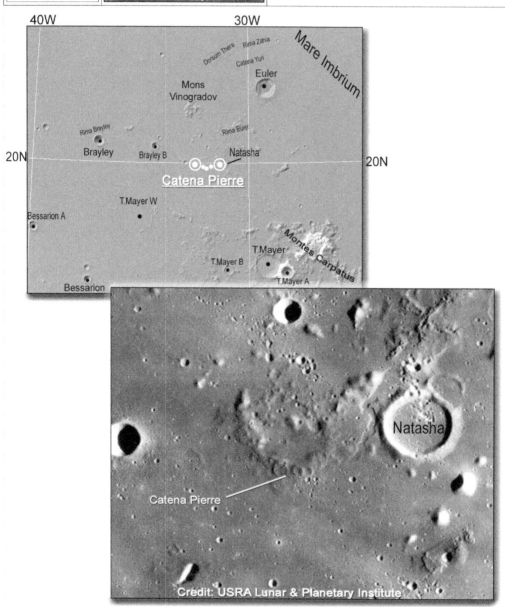

Credit: USRA Lunar & Planetary Institute

Catenae

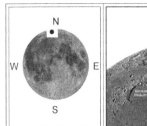

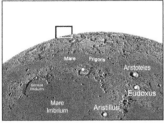

Catena Sylvester	
Lat 79.99N	Long 83.12W
Diameter	139.0Km
Notes:	

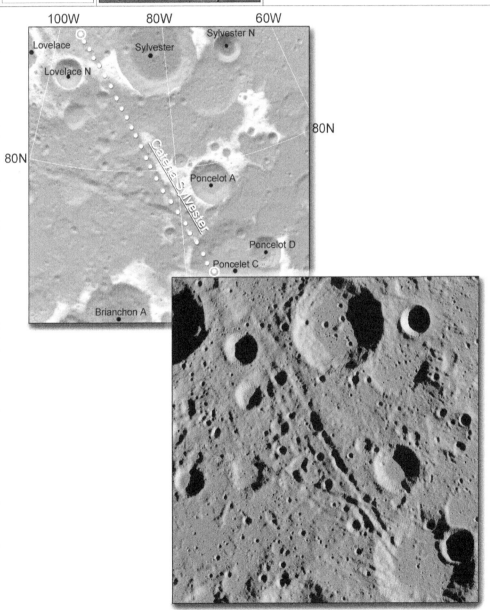

Catenae

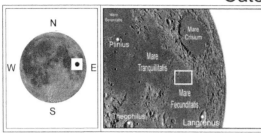

Catena Taruntius	
Lat 3.04N	Long 48.71E
Diameter	69.24Km
Notes:	

55E 50E

Taruntius

5N 5N

Taruntius T

Catena Taruntius

Taruntius B

Anville

Dorsum Cayeux

Taruntius K

Mare Fecunditatis

Taruntius H

55E

Catenae

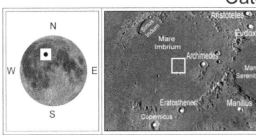

Catena Timocharis	
Lat 29.09N	Long 13.21W
Diameter	48.37Km
Notes:	

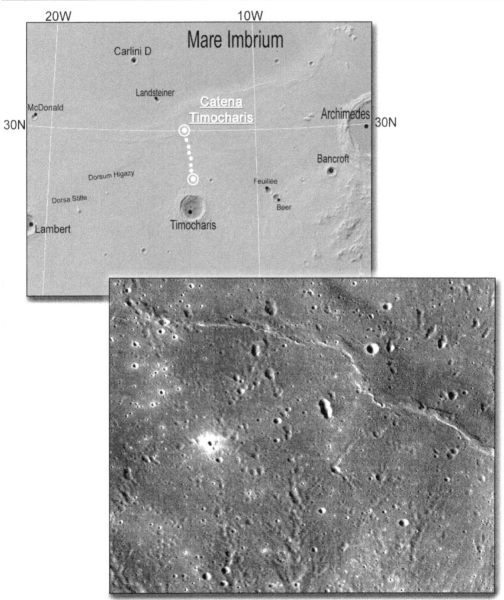

Catenae

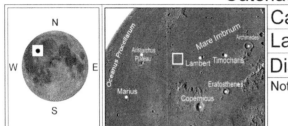

Catena Yuri	
Lat 24.41N	Long 30.38W
Diameter	4.52Km
Notes:	

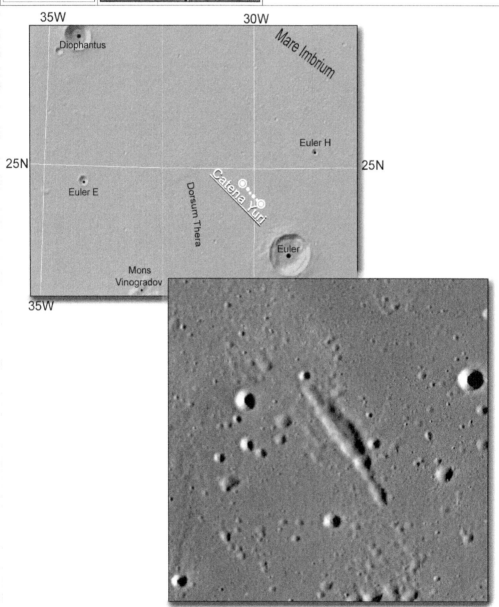

Dorsa

Dorsa

Dorsa is the plural form of the word Dorsum, which in Latin stands for 'ridge'. In the lunar context, the Dorsa appear on, and within, the dark basaltic volcanic plains known today as Maria (Seas) that filled into huge basins on the Moon. Some of the ridges appear simply linear to sinuous-like, for example Dorsa Aldrovandi, however, others, like Dorsa Lister, appear almost circular in form. In all instances however, their formation usually is in response to structural, tectonic effects within the original basins over time, leading to buckling of the surface plains above. In effect, a series of ridges form into wrinkles (they usually are called 'wrinkle ridges' or 'mare ridges'), and in most cases their orientations appear circumferential to the basins they formed in. Dorsa can be very complex in formation, which today still really aren't understood.

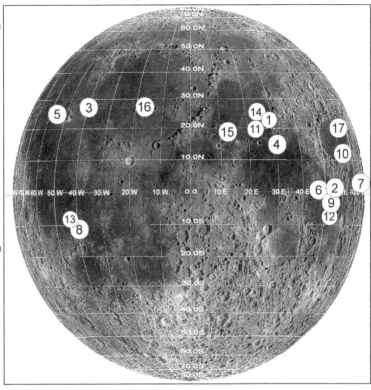

No.	Feature	No.	Feature	No.	Feature
1	Aldrovandi	7	Dana	13	Rubey
2	Andrusov	8	Ewing	14	Smirnov
3	Argand	9	Geikie	15	Sorby
4	Barlow	10	Harker	16	Tetyaev
5	Burnet	11	Lister	17	Whiston
6	Cato	12	Mawson		

Note: The above list of Dorsa represents those given in the International Astronomical Union list, but there are plenty more on the Nearside's face that may in the future be given official designations.

Dorsa

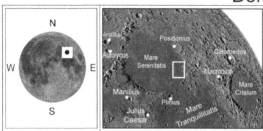

Dorsa Aldrovandi	
Lat 23.61N	Long 28.65E
Diameter	126.81Km
Notes:	

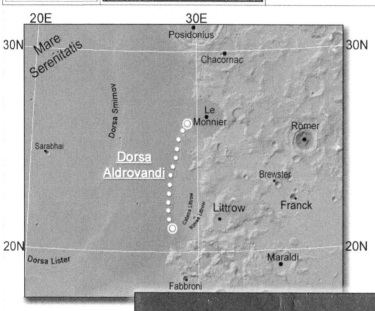

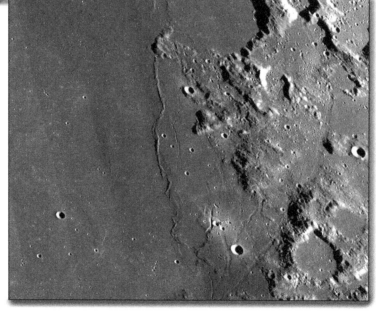

Dorsa

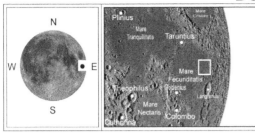

Dorsa Andrusov	
Lat 1.56S	Long 56.77E
Diameter	80.98Km
Notes:	

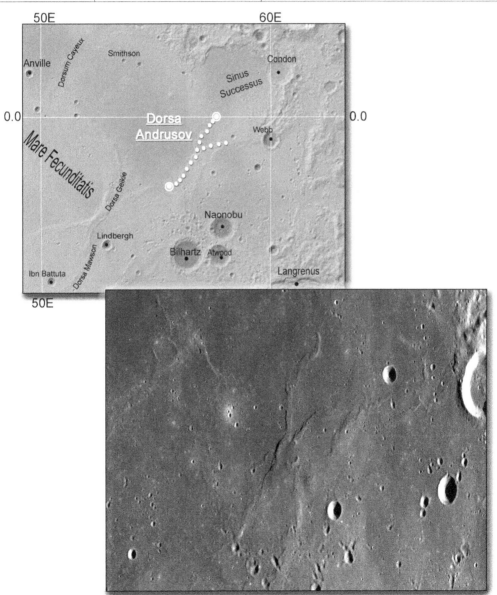

Dorsa

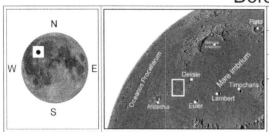

Dorsa Argand	
Lat 28.29N	Long 40.34W
Diameter	91.94Km
Notes:	

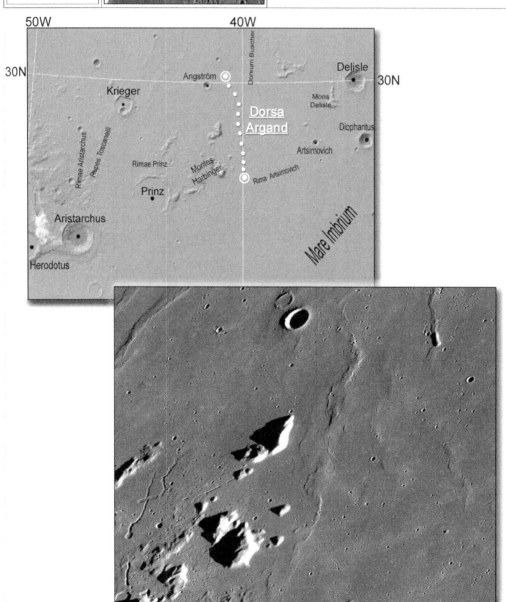

Dorsa

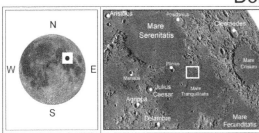

Dorsa Barlow	
Lat 14.04N	Long 30.57E
Diameter	112.57Km
Notes:	

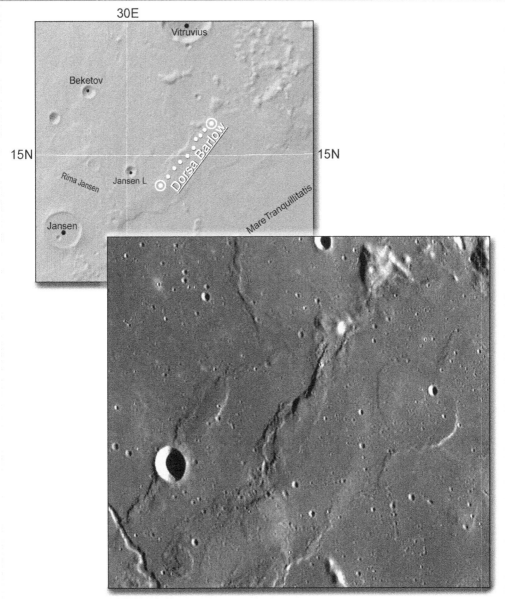

Dorsa

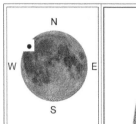

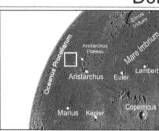

Dorsa Burnet	
Lat 26.18N	Long 56.78W
Diameter	194.99Km
Notes:	

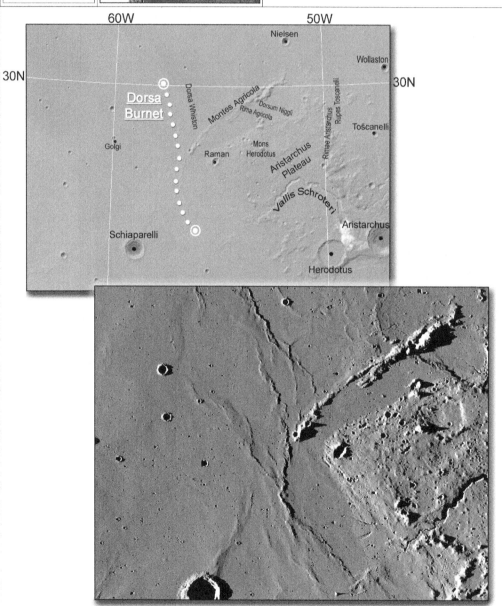

Dorsa

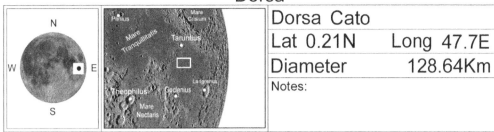

Dorsa Cato	
Lat 0.21N	Long 47.7E
Diameter	128.64Km
Notes:	

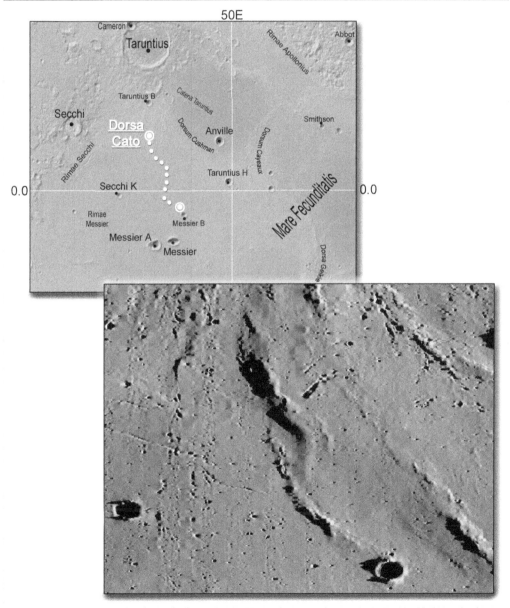

Dorsa

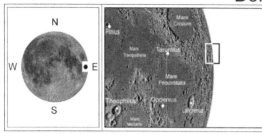

Dorsa Dana	
Lat 2.26N	Long 89.6E
Diameter	82.31Km
Notes:	

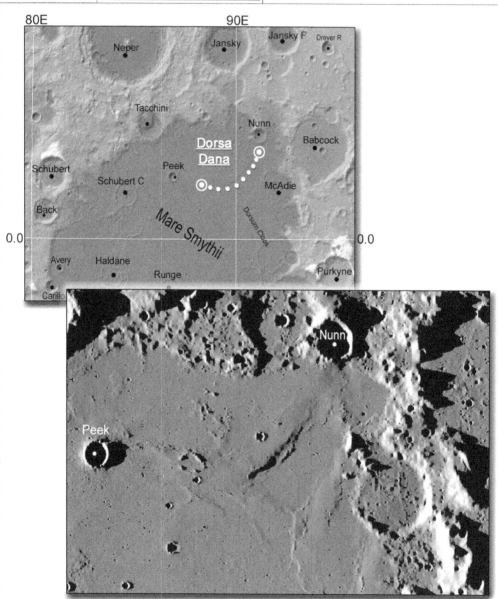

Dorsa

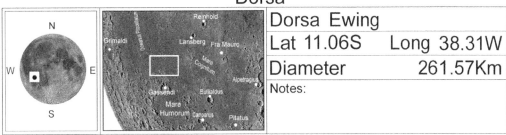

Dorsa Ewing	
Lat 11.06S	Long 38.31W
Diameter	261.57Km
Notes:	

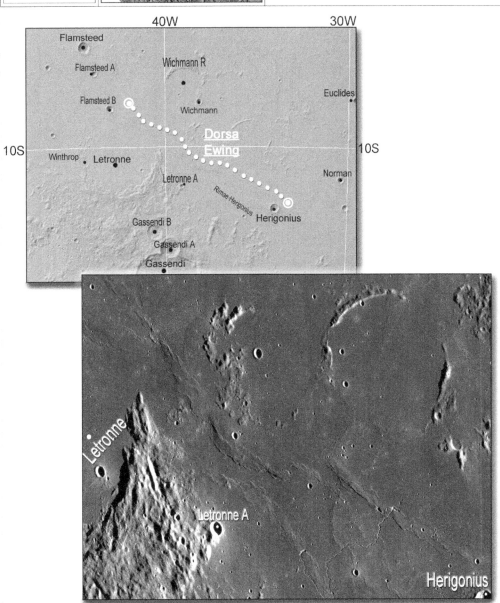

Dorsa

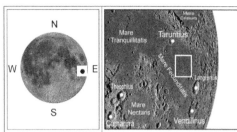

Dorsa Geikie	
Lat 4.21S	Long 52.83E
Diameter	218.35Km
Notes:	

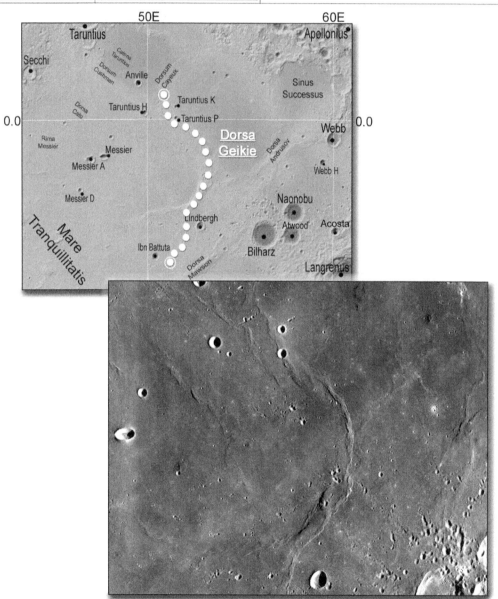

Dorsa

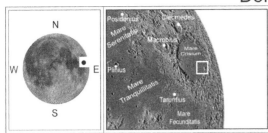

Dorsa Harker	
Lat 13.79N	Long 63.65E
Diameter	213.14Km
Notes:	

60E   65E

Mare Crisium

15N   15N

Dorsa Harker

Promontorium
Agarum

Picard Y   Fahrenheit

Condorcet J

Condorcet H   Condorcet TA

Mons Usov

Condorcet T

Dorsa

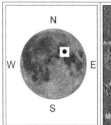

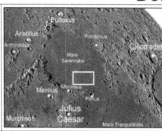

Dorsa Lister	
Lat 19.76N	Long 23.52E
Diameter	180.09Km
Notes:	

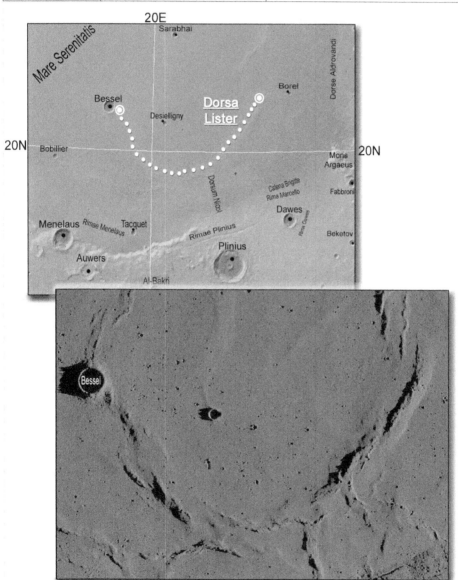

Dorsa

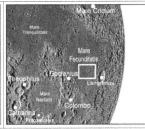

Dorsa Mawson	
Lat 7.77S	Long 52.48E
Diameter	142.71Km
Notes:	

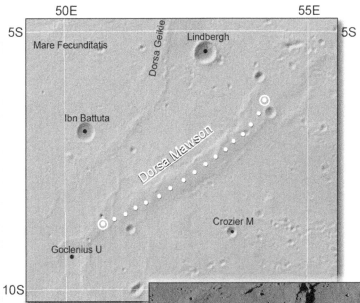

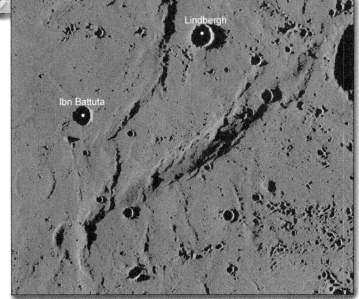

Dorsa

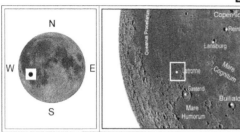

Dorsa Rubey	
Lat 9.88S	Long 42.36W
Diameter	100.64Km
Notes:	

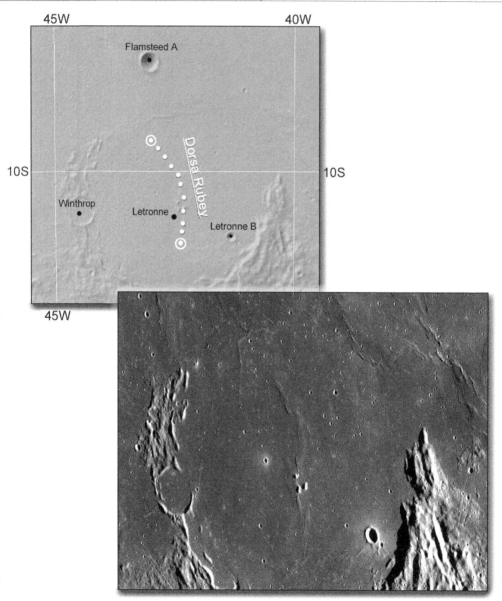

Dorsa

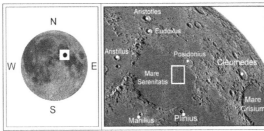

Dorsa Smirnov	
Lat 26.41N	Long 25.53E
Diameter	221.78Km
Notes:	

31

Dorsa

Dorsa Sorby	
Lat 18.65N	Long 13.7E
Diameter	76.03Km
Notes:	

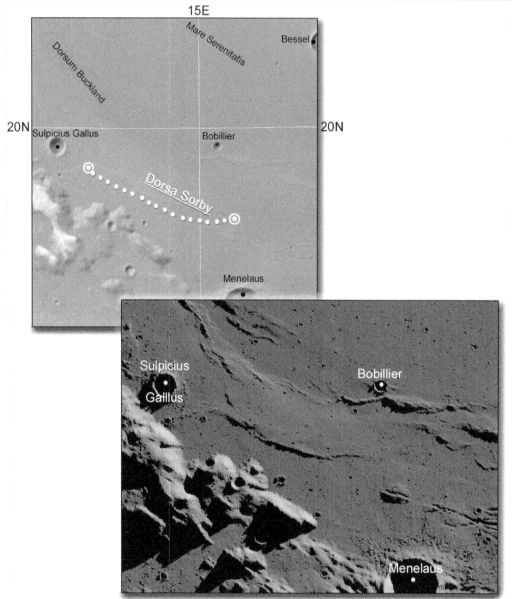

Dorsa

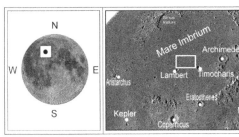

Dorsa Stille	
Lat 26.91N	Long 18.85W
Diameter	66.32Km
Notes:	

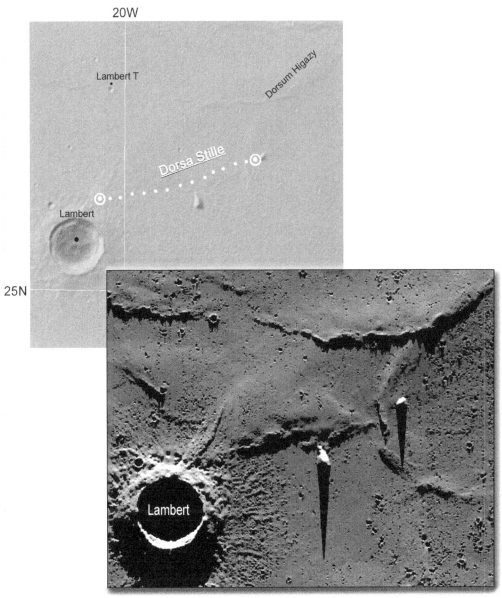

Dorsa

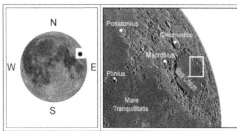

Dorsa Tetyaev	
Lat 20.02N	Long 64.06E
Diameter	187.53Km
Notes:	

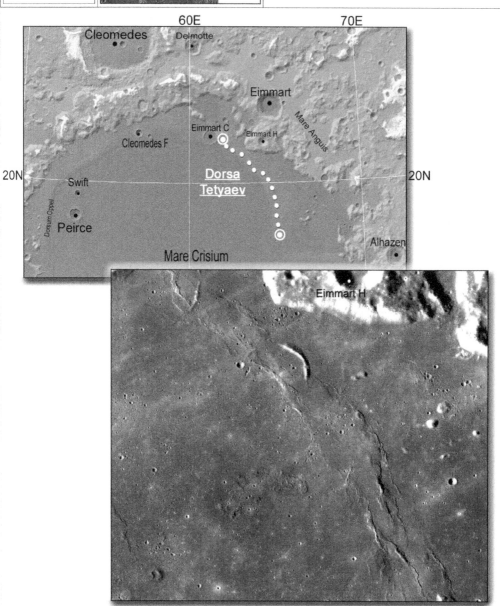

Dorsa

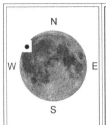

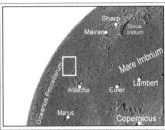

Dorsa Whiston	
Lat 29.77N	Long 56.96W
Diameter	138.93Km
Notes:	

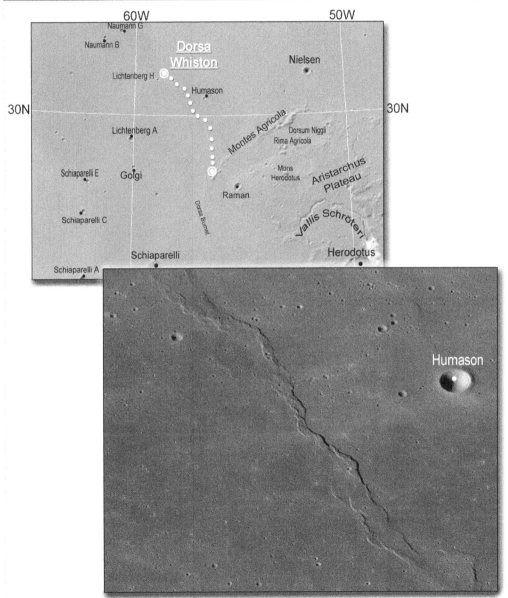

Dorsum

Dorsum

Dorsum on the moon's surface refers to ridges that formed (usually in basins and in the plains of dark basaltic material that fill them) in response to structural shifts underneath. Dorsum usually refers to a single ridge (the plural is Dorsa - see the Dorsa pages), however, when looking at both types it's not so easy to distinguish 'a single ridge' to 'a group of ridges' as nearly all show similar distorted and buckled effects in each.

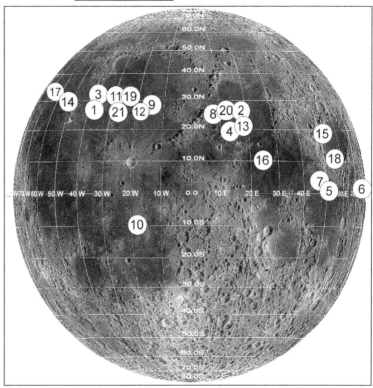

No.	Feature	No.	Feature	No.	Feature
1	Arduino	8	Gast	15	Oppel
2	Azara	9	Grabau	16	Owen
3	Bucher	10	Geuttard	17	Scilla
4	Buckland	11	Heim	18	Termier
5	Cayeux	12	Higazy	19	Thera
6	Cloos	13	Nicol	20	Von Cotta
7	Cushman	14	Niggli	21	Zirkel

Note: The above list of Dorsum represents those given in the International Astronomical Union list, but there are plenty more on the Nearside's face that may in the future be given official designations.

Dorsum

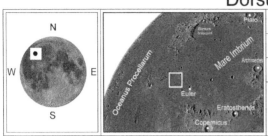

Dorsum Arduino	
Lat 24.77N	Long 36.27W
Diameter	99.73Km
Notes:	

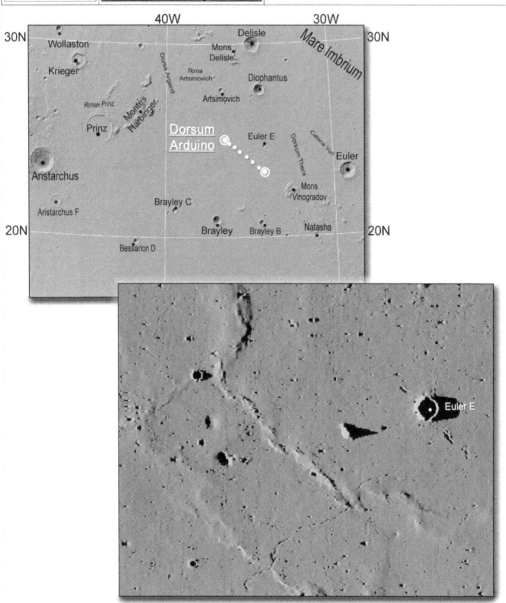

Dorsum

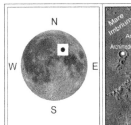

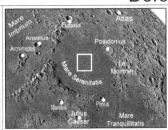

Dorsum Azara	
Lat 26.86N	Long 19.17E
Diameter	103.2Km
Notes:	

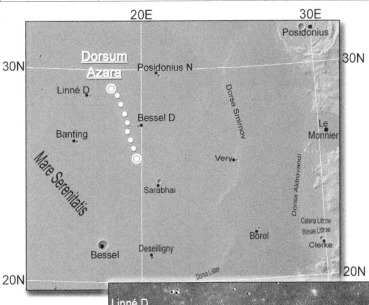

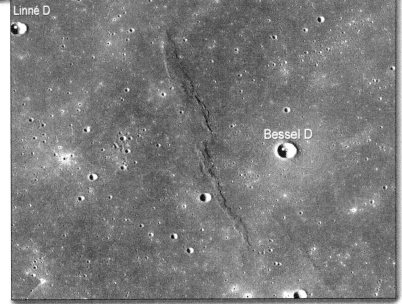

Dorsum

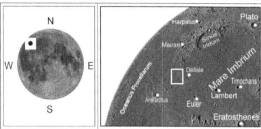

Dorsum Bucher	
Lat 30.76N	Long 39.55W
Diameter	84.65Km
Notes:	

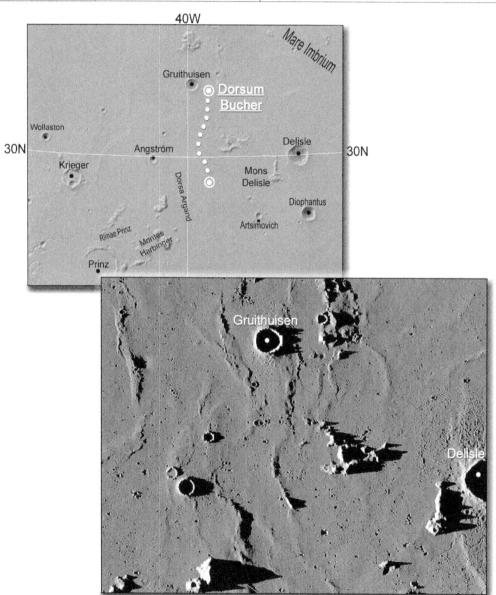

Dorsum

Dorsum Buckland	
Lat 19.43N	Long 14.3E
Diameter	369.13Km
Notes:	

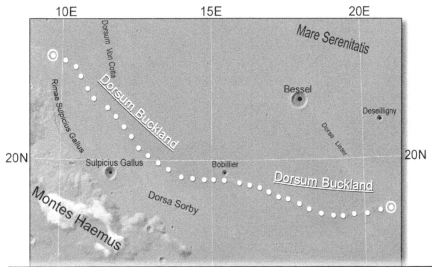

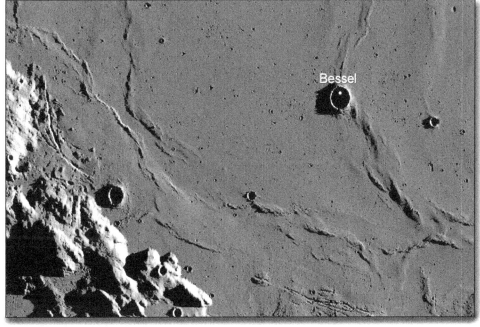

Dorsum

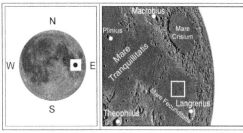

Dorsum Cayeux	
Lat 0.76N	Long 51.22E
Diameter	95.14Km
Notes:	

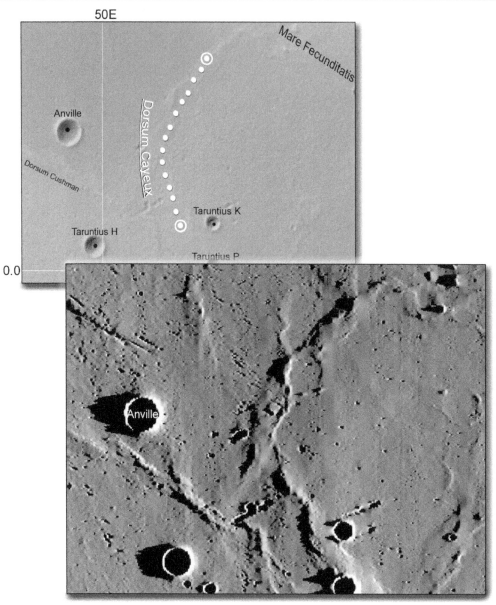

Dorsum

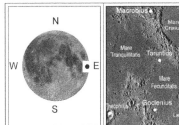

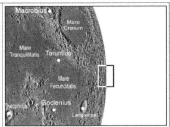

Dorsum Cloos	
Lat 1.15N	Long 90.41E
Diameter	103.09Km
Notes:	

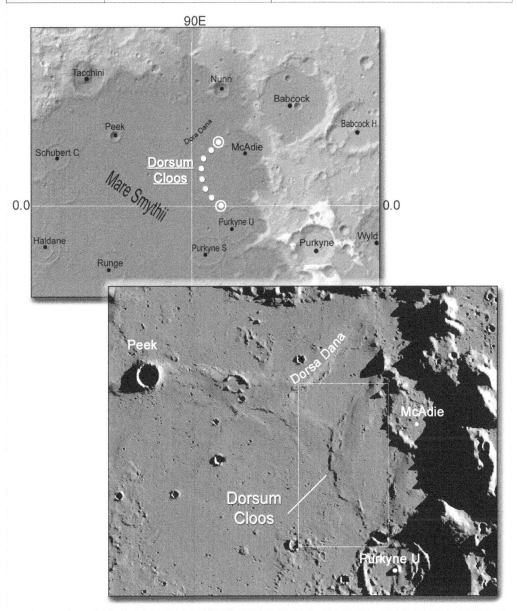

Dorsum

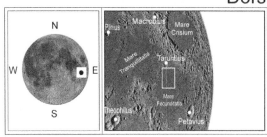

Dorsum Cushman	
Lat 1.42N	Long 49.19E
Diameter	85.65Km
Notes:	

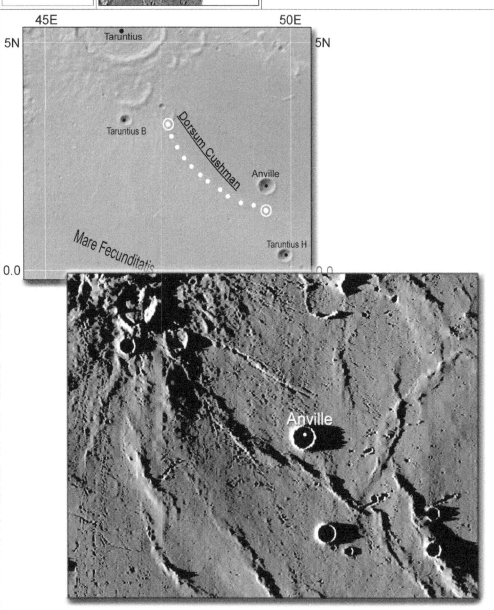

Dorsum

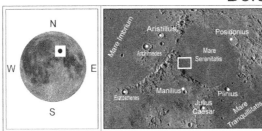

Dorsum Gast	
Lat 24.38N	Long 8.71E
Diameter	64.87Km
Notes:	

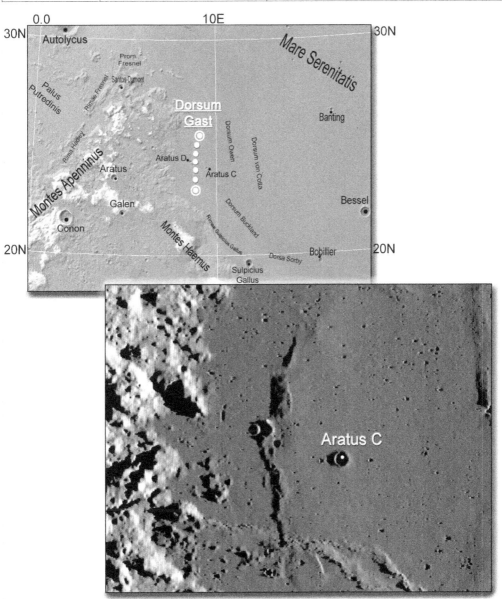

Dorsum

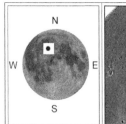

Dorsum Grabau	
Lat 29.76N	Long 14.19W
Diameter	123.69Km
Notes:	

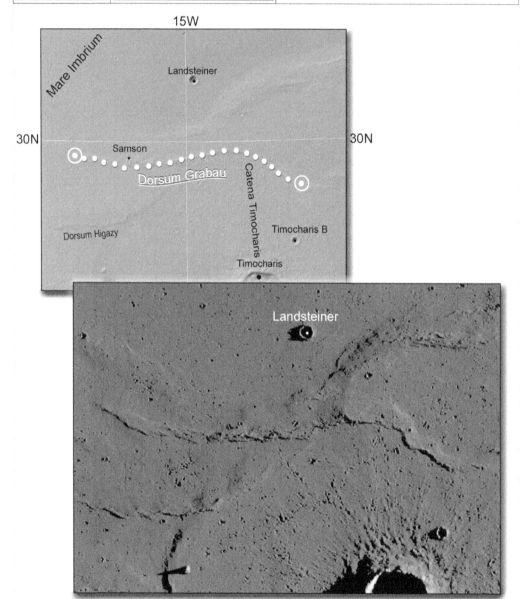

Dorsum

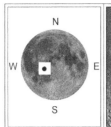

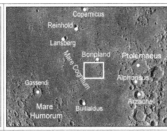

Dorsum Guettard	
Lat 9.92S	Long 18.26W
Diameter	40.46Km
Notes:	

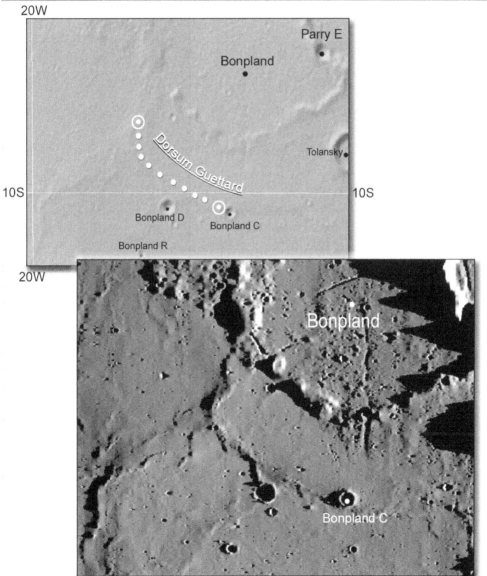

Dorsum

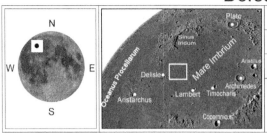

Dorsum Heim	
Lat 32.2N	Long 29.83W
Diameter	146.79Km
Notes:	

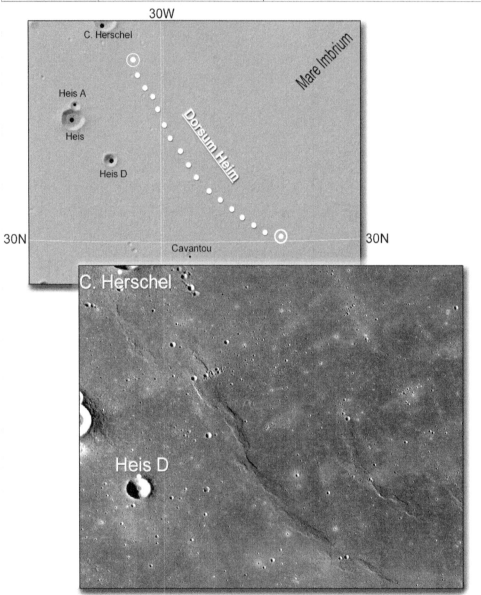

Dorsum

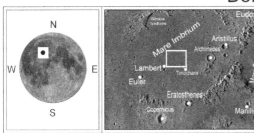

Dorsum Higazy	
Lat 27.93N	Long 17.47W
Diameter	63.1Km
Notes:	

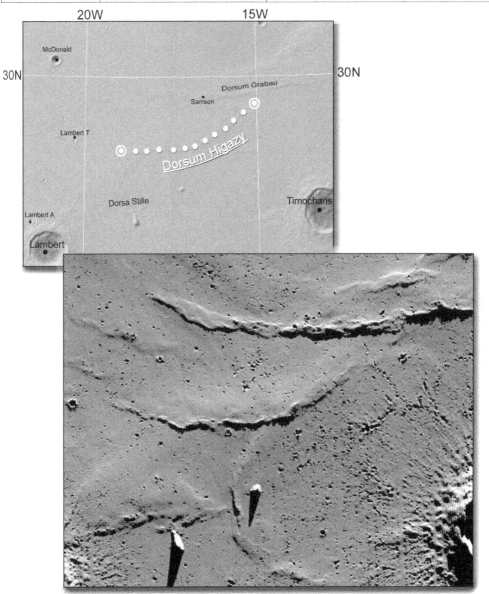

Dorsum

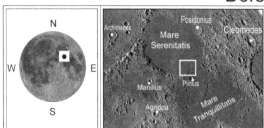

Dorsum Nicol	
Lat 18.32N	Long 22.66E
Diameter	43.74Km
Notes:	

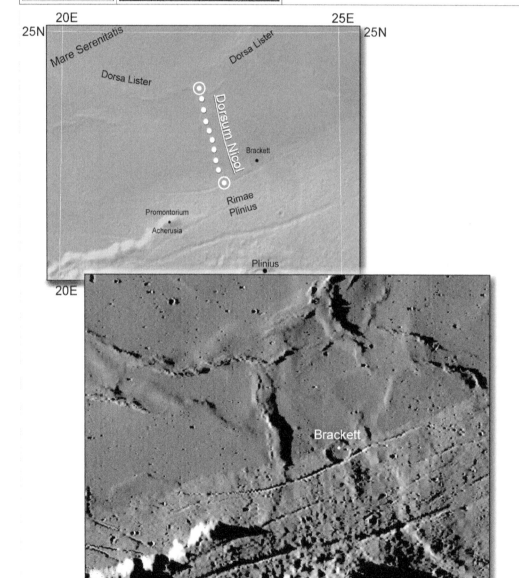

Dorsum

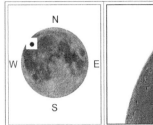

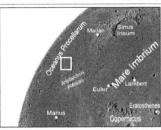

Dorsum Niggli	
Lat 29.01N Long 52.28W	
Diameter	47.75Km
Notes:	

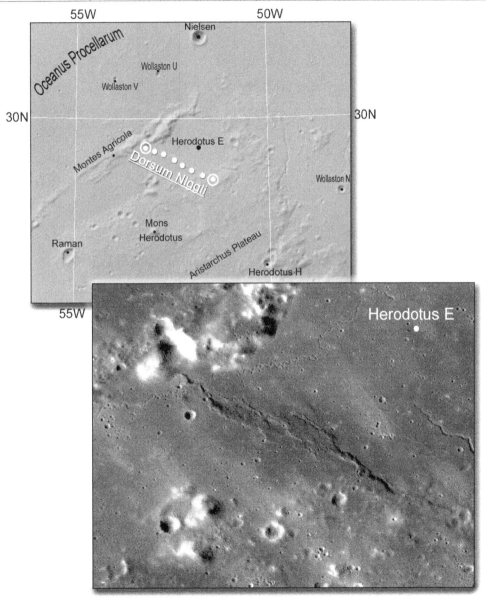

51

Dorsum

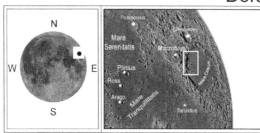

Dorsum Oppel	
Lat 19.31N	Long 52.09E
Diameter	297.62 Km
Notes:	

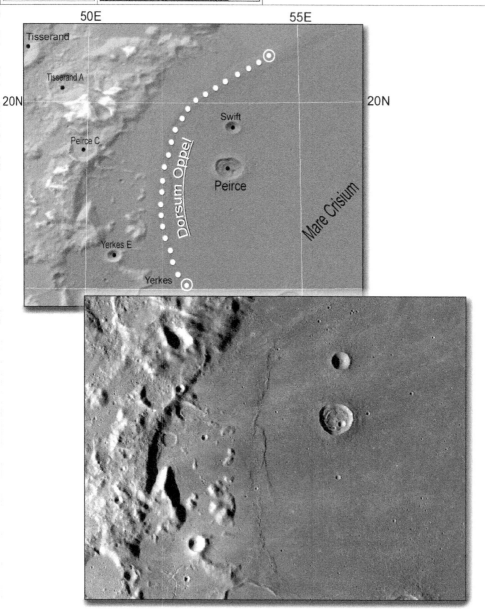

Dorsum

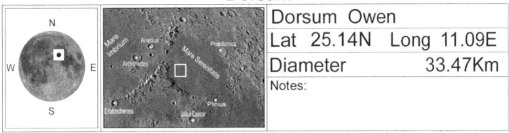

Dorsum Owen	
Lat 25.14N Long 11.09E	
Diameter	33.47Km
Notes:	

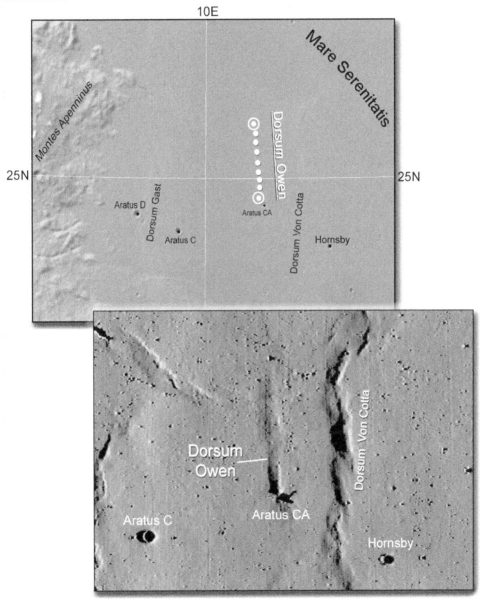

Dorsum

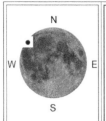

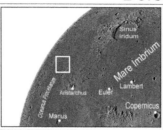

Dorsum Scilla		
Lat 32.34N	Long	60.0W
Diameter		107.52Km
Notes:		

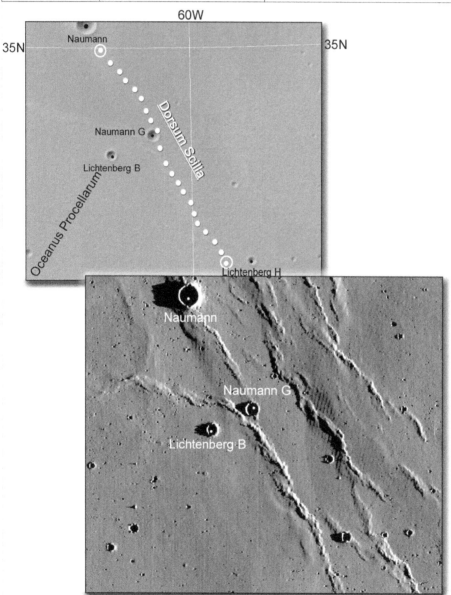

Dorsum

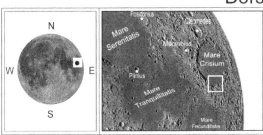

Dorsum Termier	
Lat 11.63N	Long 57.15E
Diameter	89.65Km
Notes:	

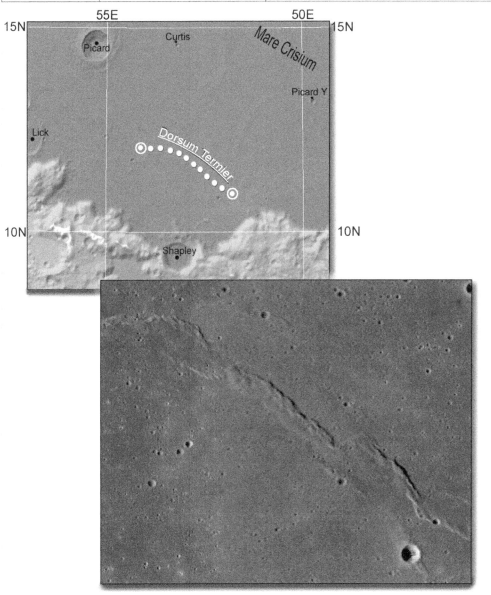

Dorsum

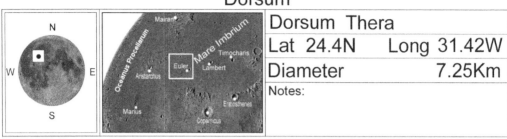

Dorsum Thera	
Lat 24.4N	Long 31.42W
Diameter	7.25Km
Notes:	

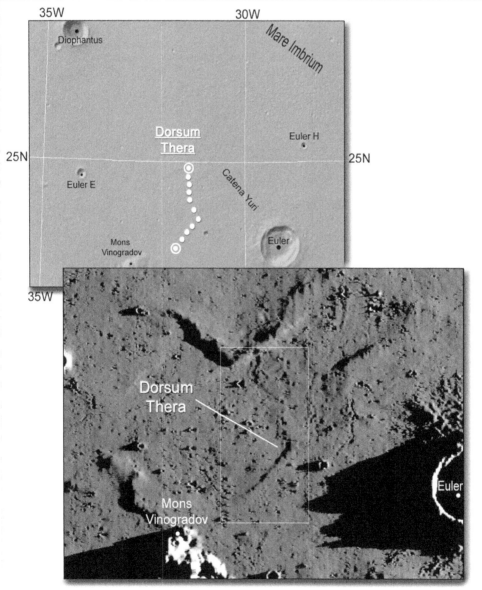

Dorsum

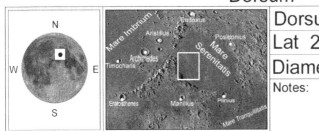

Dorsum Von Cotta	
Lat 23.6N	Long 11.95E
Diameter	183.06Km
Notes:	

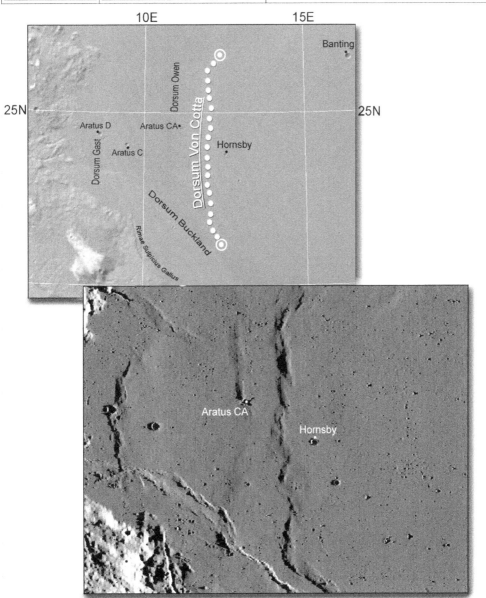

Dorsum

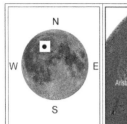

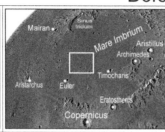

Dorsum Zirkel	
Lat 29.55N	Long 24.82W
Diameter	195.22Km
Notes:	

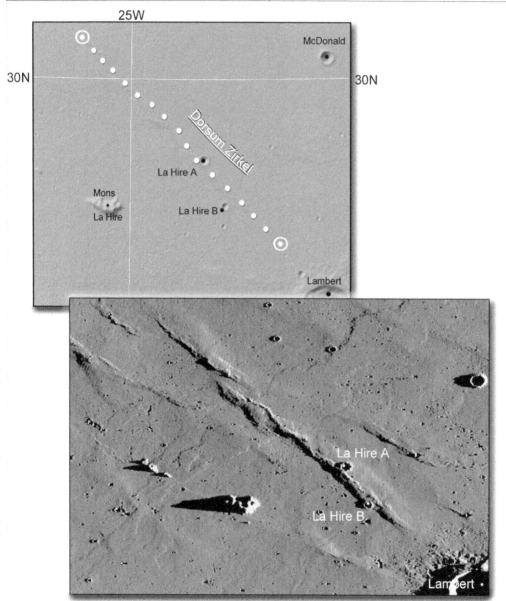

Lacus

Lacus

Lacus generally translates from the Latin to 'lake', or 'opening'. In the lunar context it usually refers to small isolated patches of mare where their associated basaltic materials have either extruded onto the surface through local faults, or form part of a much larger off-branch from a main mare nearby.

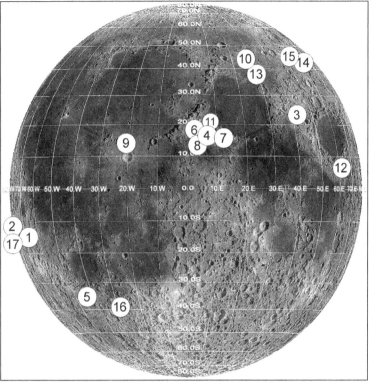

No.	Feature			No.	Feature		
1	Aestatis	-	Lake of Summer	10	Mortis	-	Lake of Death
2	Autumni	-	Lake of Autumn	11	Odii	-	Lake of Hatred
3	Bonitatis	-	Lake of Goodness	12	Perseverantiae	-	Lake of Perseverance
4	Doloris	-	Lake of Sorrow	13	Somniorum	-	Lake of Dreams
5	Excellentiae	-	Lake of Excellence	14	Spei	-	Lake of Hope
6	Felicitatis	-	Lake of Happiness	15	Temporis	-	Lake of Time
7	Gaudii	-	Lake of Joy	16	Timoris	-	Lake of Fear
8	Hiemalis	-	Wintry Lake	17	Veris	-	Lake of Spring
9	Lenitatis	-	Lake of Softness				

Note: The above list of Lacus represents those given in the International Astronomical Union list, which may grow in the near future as more are added with official designations.

Lacus

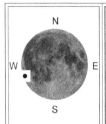
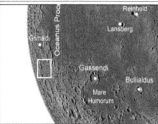

Lacus Aestatis	
Lat 14.83S	Long 68.57W
Diameter	86.39Km
Notes:	

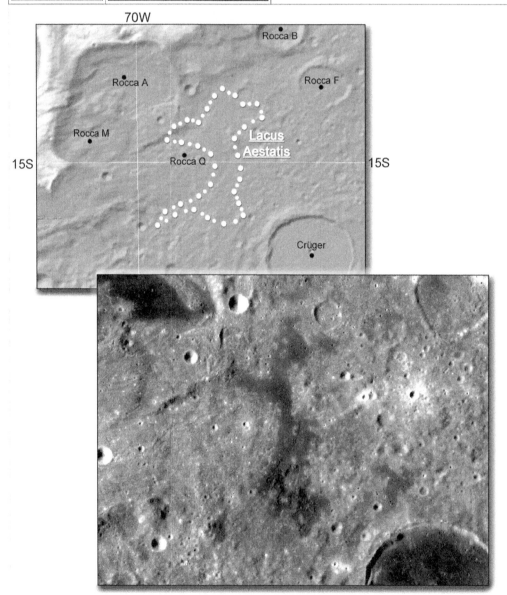

Lacus

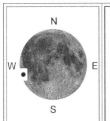

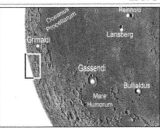

Lacus Autumni	
Lat 11.81S	Long 83.17W
Diameter	195.65Km
Notes:	

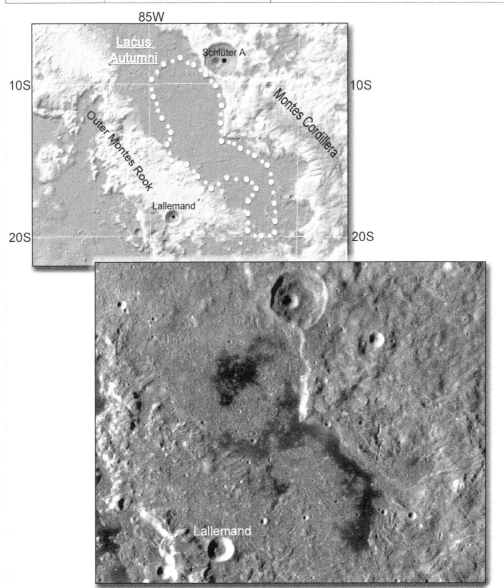

Lacus

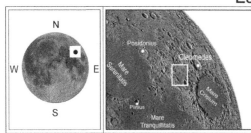

Lacus Bonitatis	
Lat 23.18N	Long 44.32E
Diameter	122.1Km
Notes:	

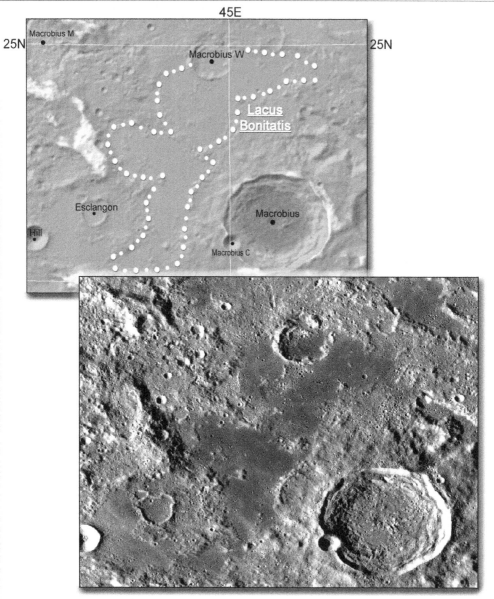

Lacus

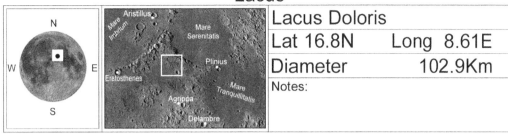

Lacus Doloris	
Lat 16.8N	Long 8.61E
Diameter	102.9Km
Notes:	

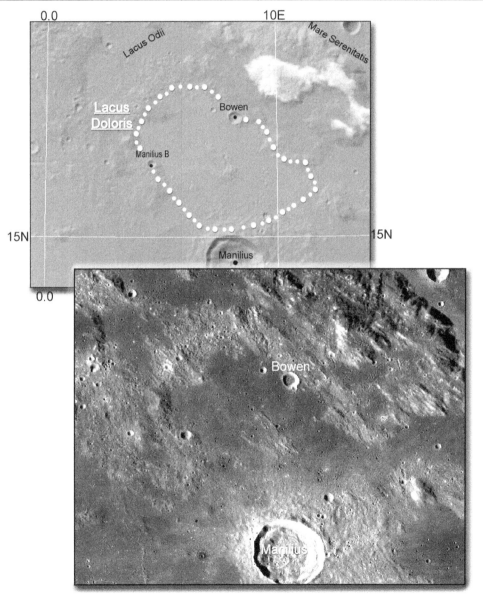

Lacus

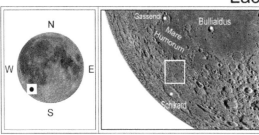

Lacus Excellentiae	
Lat 35.65S	Long 43.58W
Diameter	197.74Km
Notes:	

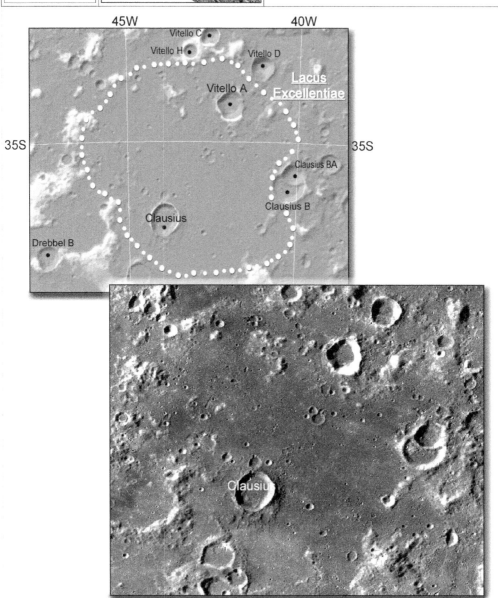

Lacus

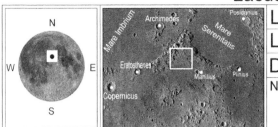

Lacus Felicitatis	
Lat 18.52N	Long 5.36E
Diameter	98.48Km
Notes:	

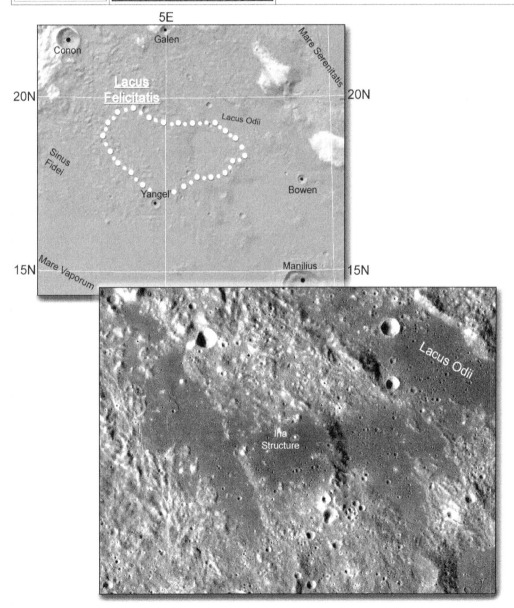

Lacus

		Lacus Gaudii	
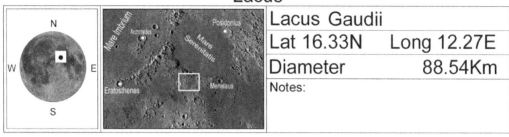		Lat 16.33N	Long 12.27E
		Diameter	88.54Km
		Notes:	

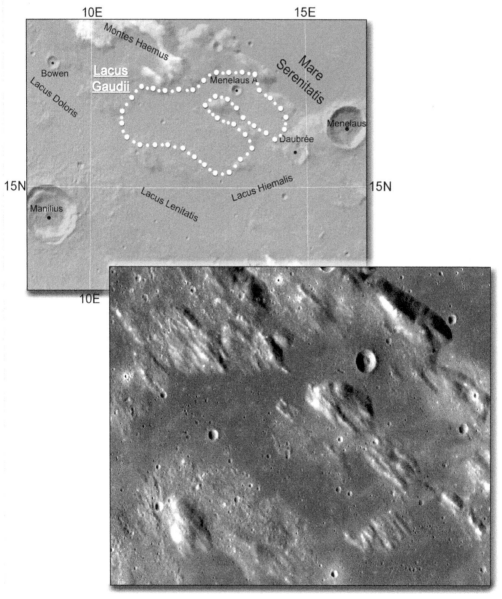

Lacus

Lacus Hiemalis	
Lat 15.01N	Long 13.97E
Diameter	48.04Km
Notes:	

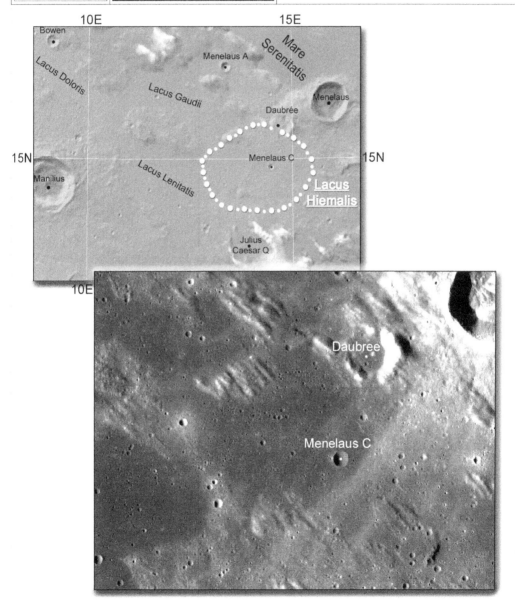

Lacus

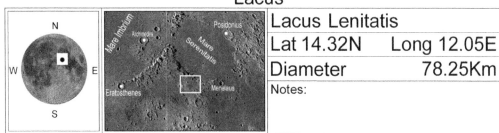

Lacus Lenitatis	
Lat 14.32N	Long 12.05E
Diameter	78.25Km
Notes:	

10E   15E

Bowen

Mare Imbrium

Archimedes

Posidonius

Mare Serenitatis

Eratosthenes

Menelaus

Mare Serenitatis

Menelaus A

Lacus Doloris

Lacus Gaudii

Menelaus

Daubrée

15N   15N

Manilius

Menelaus C

Lacus Hiemalis

Lacus Lenitatis

Julius
Caesar Q

10E

Lacus

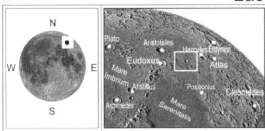

Lacus Mortis	
Lat 45.13N	Long 27.32E
Diameter	158.78Km
Notes:	

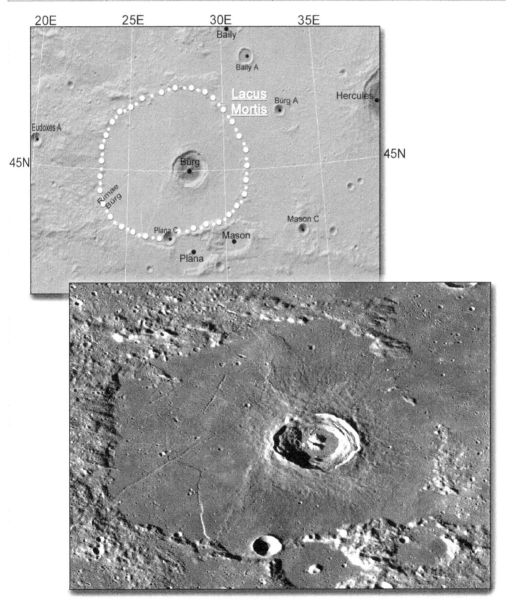

Lacus

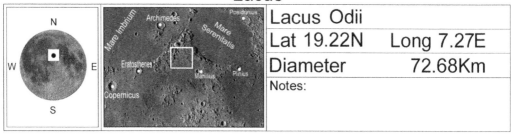

Lacus Odii	
Lat 19.22N	Long 7.27E
Diameter	72.68Km
Notes:	

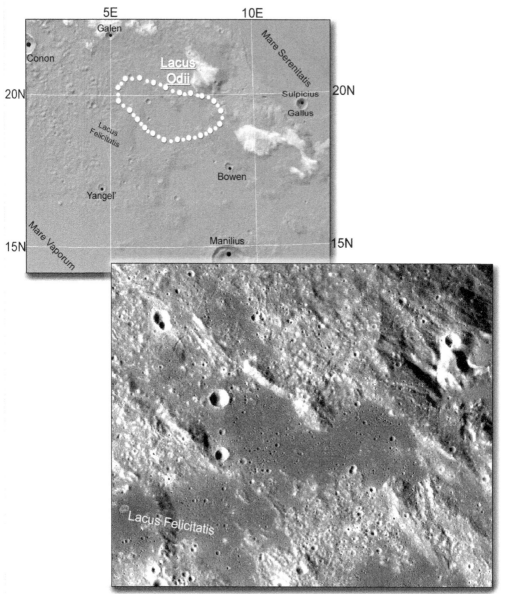

Lacus

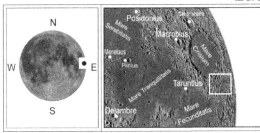

Lacus Perseverantiae	
Lat 7.84N	Long 61.93E
Diameter	70.64Km
Notes:	

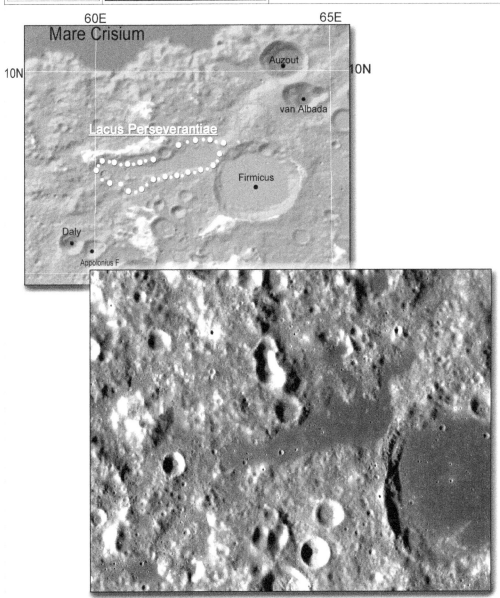

Lacus

Lacus Somniorum	
Lat 37.56N	Long 30.8E
Diameter	424.76Km
Notes:	

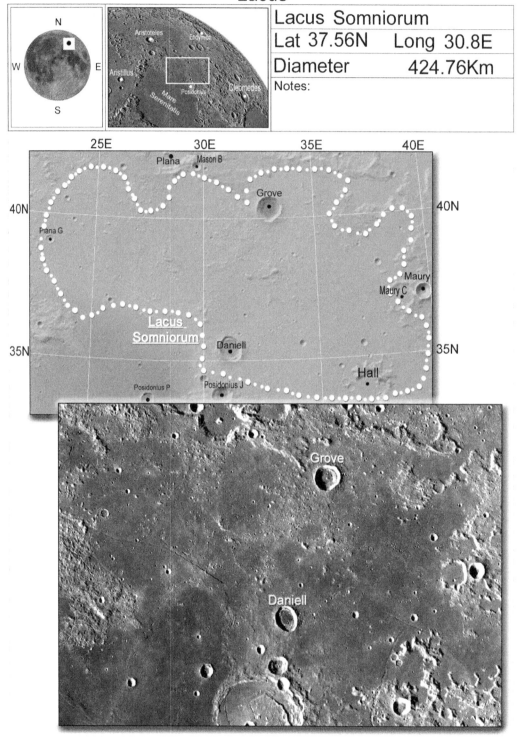

Lacus

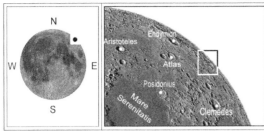

Lacus Spei	
Lat 43.46N	Long 65.2E
Diameter	76.67Km
Notes:	

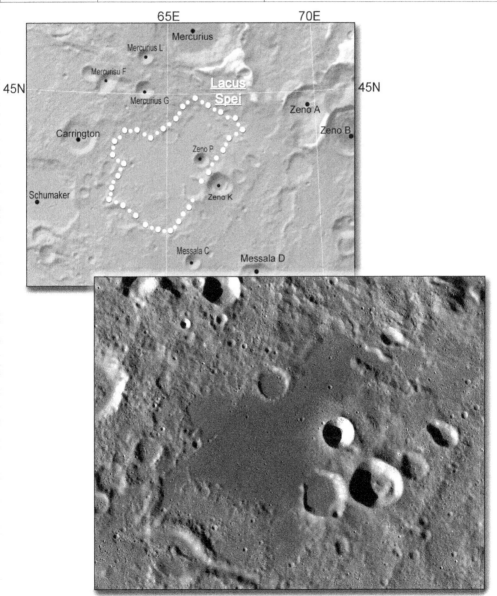

Lacus

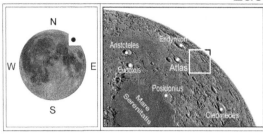

Lacus Temporis	
Lat 46.77N	Long 56.21E
Diameter	205.3Km
Notes:	

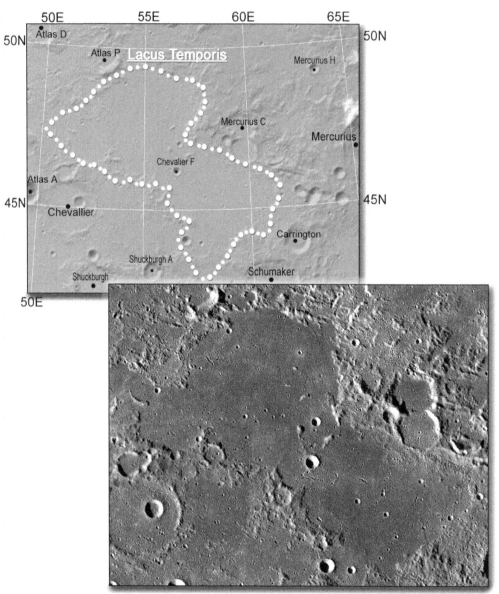

Lacus

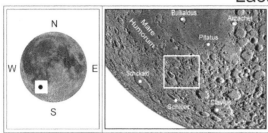

Lacus Timoris	
Lat 39.42S	Long 27.95W
Diameter	153.65Km
Notes:	

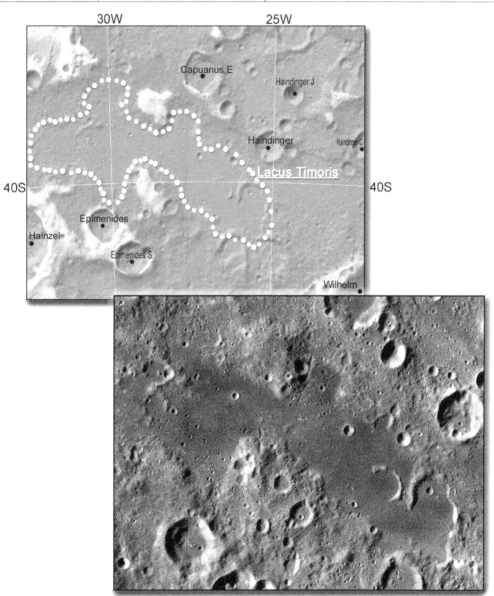

75

Lacus

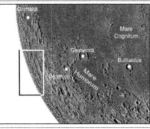

Lacus Veris	
Lat 16.48S	Long 85.91W
Diameter	382.88Km
Notes:	

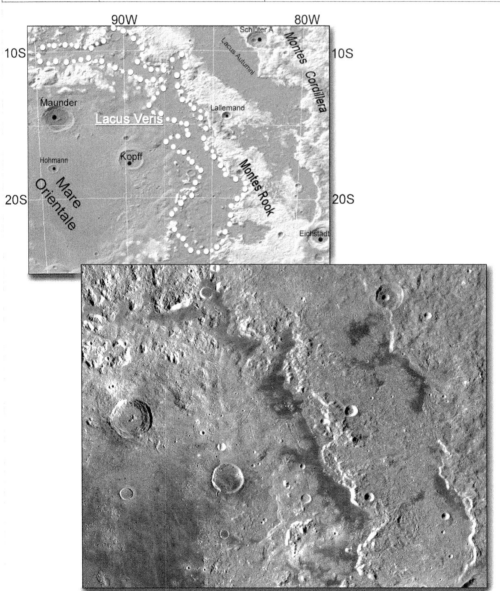

Maria

Maria

One of the most instant, visual features on the moon's Nearside must surely be the *Maria*, which in Latin stands for 'seas'. In most instances the Maria predominantly form huge basins (produced by extremely large impactors in the early forming Moon) that have been infilled with basaltic material over time. This material is believed to have leaked up through major fractures created initially in the target rock, which then filled into the depression formed (infill, in most cases, may have formed as a series of basaltic plains - each occurring within millions of years in between). Estimates put most of the basins formed within the 4.1 - 3.85 billion year old time frame (the Moon is some ~ 4.6 billions of

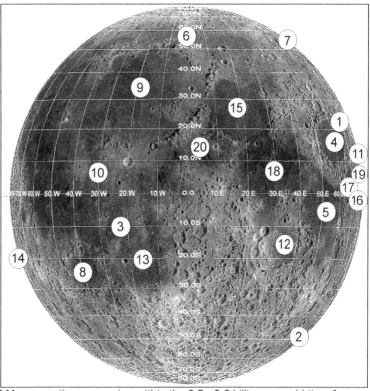

years old), with majority of Mare eruptions occurring within the 3.5 - 3.0 billion year old time frame (though, some lunar samples returned by the Apollo missions show both younger and older times, too).

No.	Feature		No.	Feature	
1	Anguis	- Serpent Sea	11	Marginis	- Sea of the Edge
2	Australe	- Southern Sea	12	Nectaris	- Sea of Nectar
3	Cognitum	- Known Sea	13	Nubium	- Sea of Clouds
4	Crisium	- Sea of Crisis	14	Orientale	- Eastern Sea
5	Fecunditatis	- Sea of Fecundity	15	Serenitatis	- Sea of Serenity
6	Frigoris	- Sea of Cold	16	Smythii	- Smyth, William Henry
7	Humboldtianum	- Humboldt, Alexander von	17	Spumans	- Foaming Sea
8	Humorum	- Sea of Moisture	18	Tranquillitatis	- Sea of Tranquillity
9	Imbrium	- Sea of Showers	19	Undarum	- Sea of Waves
10	Insularum	- Sea of Islands	20	Vaporum	- Sea of Vapors

Note: The above list of Maria represents those given in the International Astronomical Union list

Maria

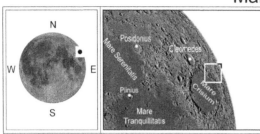

Mare Anguis	
Lat 22.43N	Long 67.58E
Diameter	145.99Km
Notes:	

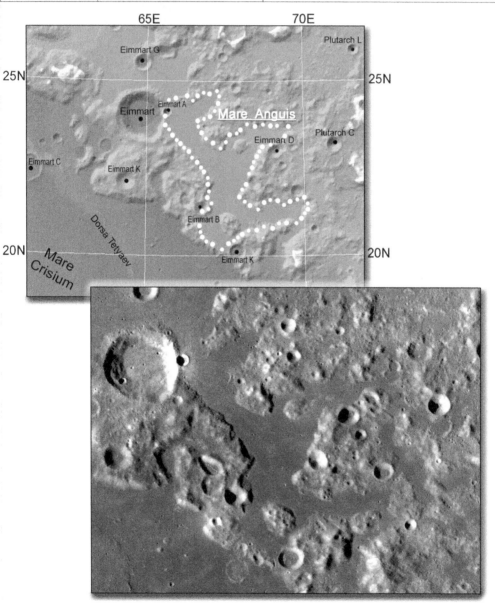

Maria

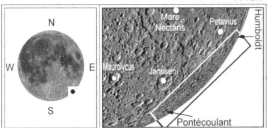

Mare Australe	
Lat 47.77S	Long 91.99E
Diameter	996.84Km
Notes:	

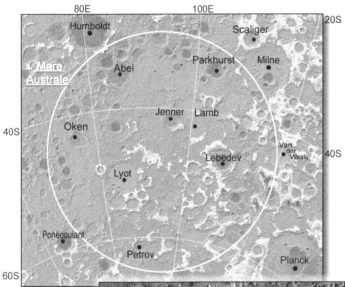

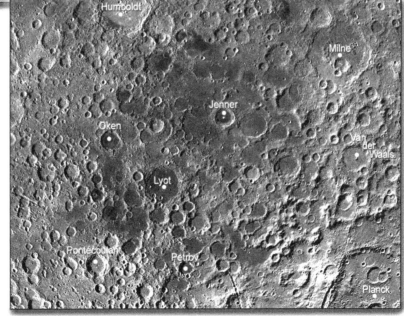

Maria

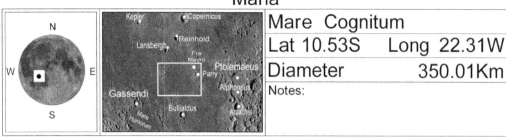

Mare Cognitum	
Lat 10.53S	Long 22.31W
Diameter	350.01Km
Notes:	

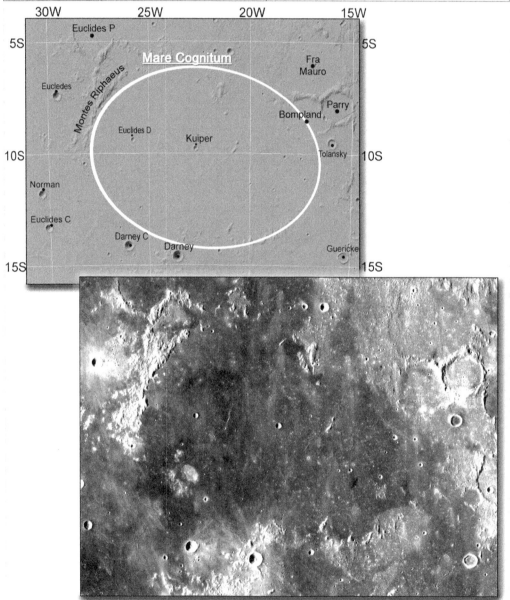

Maria

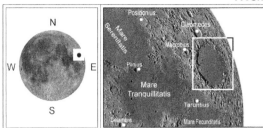

Mare Crisium	
Lat 16.18N	Long 59.1E
Diameter	555.92Km
Notes:	

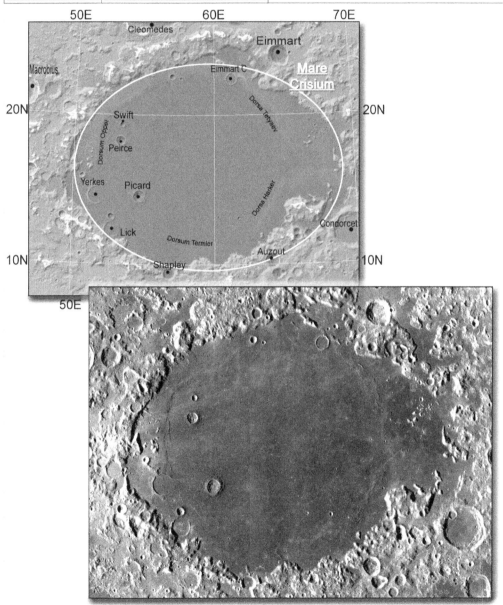

Maria

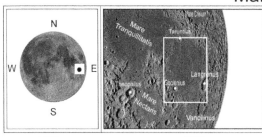

Mare Fecunditatis	
Lat 7.83S	Long 53.67E
Diameter	840.35Km
Notes:	

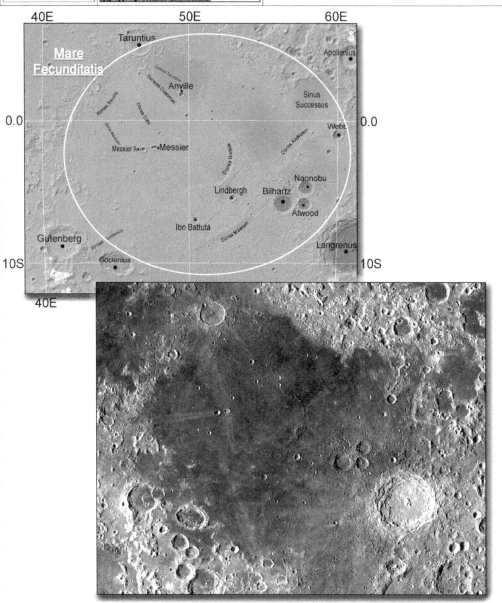

Maria

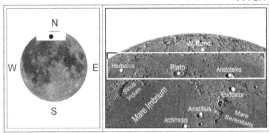

Mare Frigoris	
Lat 57.59N	Long 0.01W
Diameter	1446.41Km
Notes:	

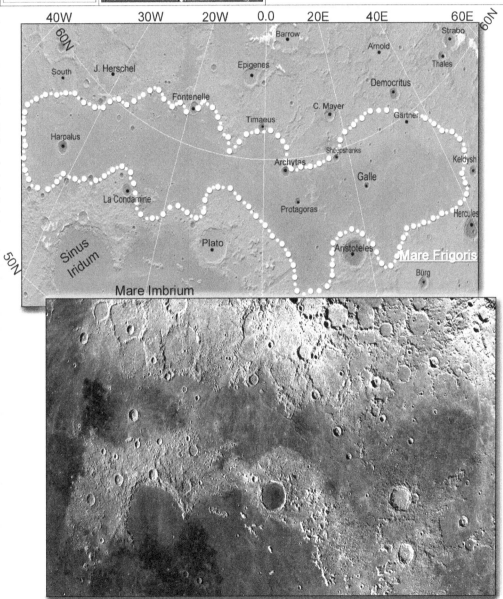

Maria

Mare Humboldtianum	
Lat 56.92N	Long 81.54E
Diameter	230.78Km
Notes:	

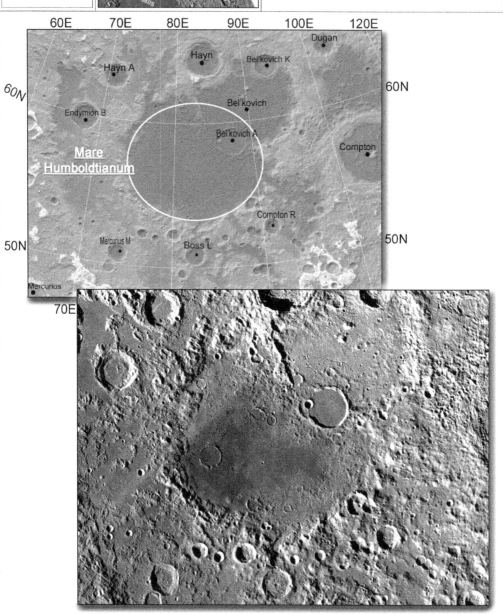

Maria

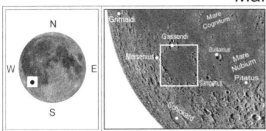

Mare Humorum	
Lat 24.48S	Long 38.57W
Diameter	419.67Km
Notes:	

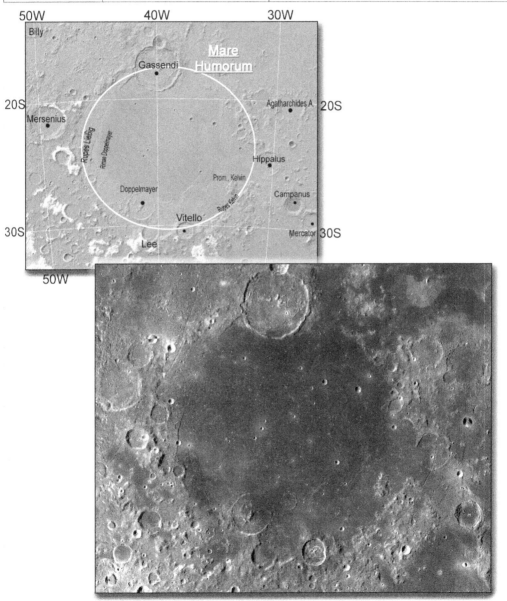

Maria

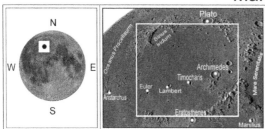

Mare Imbrium	
Lat 34.72N	Long 14.91W
Diameter	1145.53Km
Notes:	

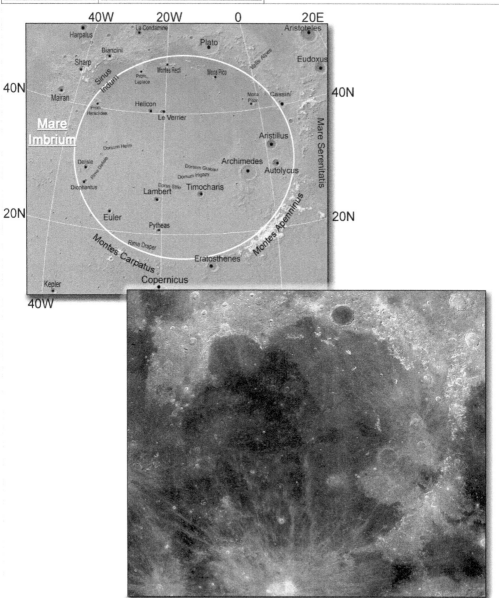

Maria

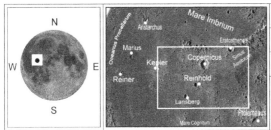

Mare Insularum	
Lat 7.79N	Long 30.64W
Diameter	511.93Km
Notes:	

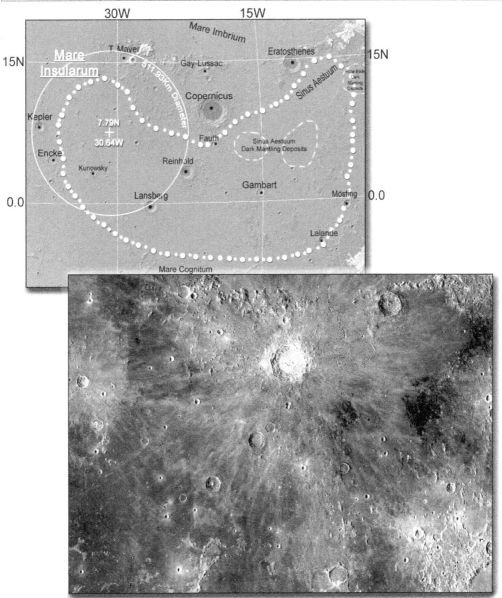

Maria

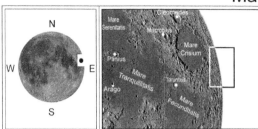

Mare Marginis	
Lat 12.7N	Long 86.52E
Diameter	357.63Km
Notes:	

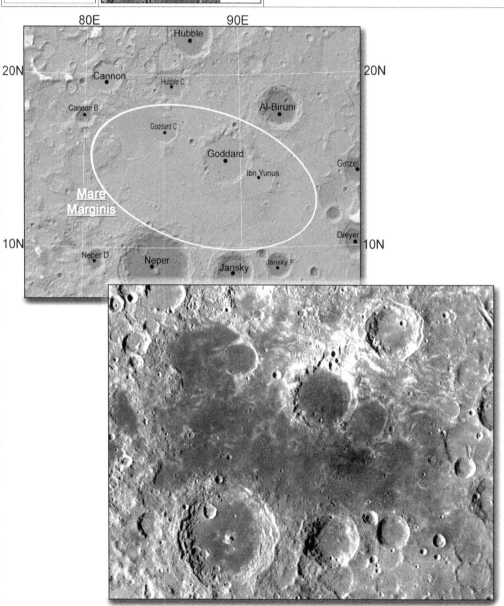

Maria

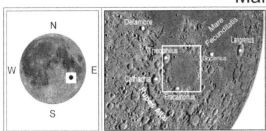

Mare Nectaris	
Lat 15.19S	Long 34.6E
Diameter	339.39Km
Notes:	

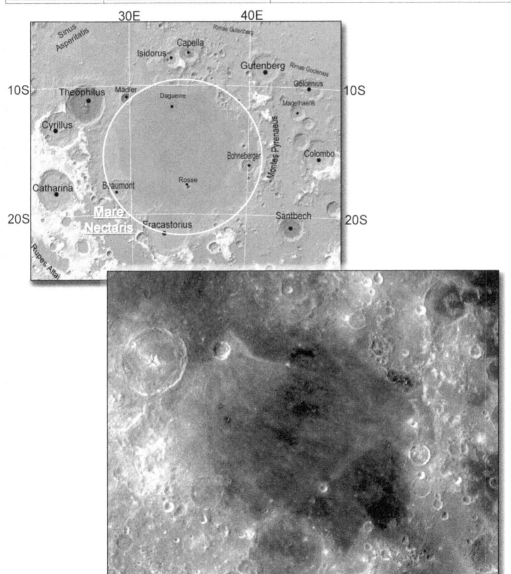

Maria

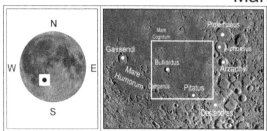

Mare Nubium	
Lat 20.59S	Long 17.29W
Diameter	714.5Km
Notes:	

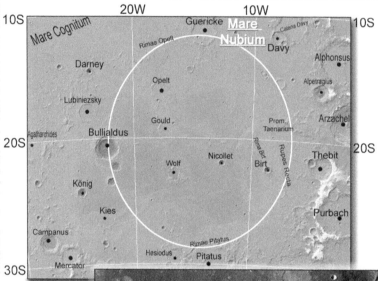

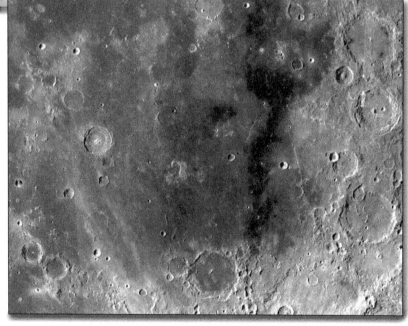

Maria

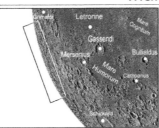

Mare Orientale	
Lat 19.87S	Long 94.67W
Diameter	294.16Km
Notes:	

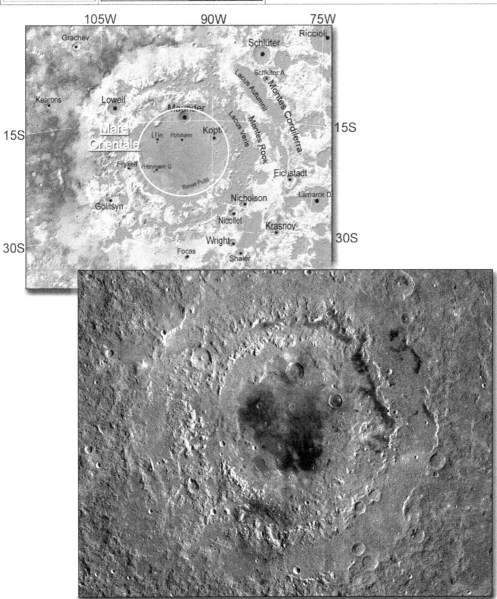

Maria

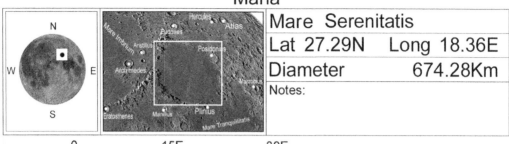

Mare Serenitatis	
Lat 27.29N	Long 18.36E
Diameter	674.28Km
Notes:	

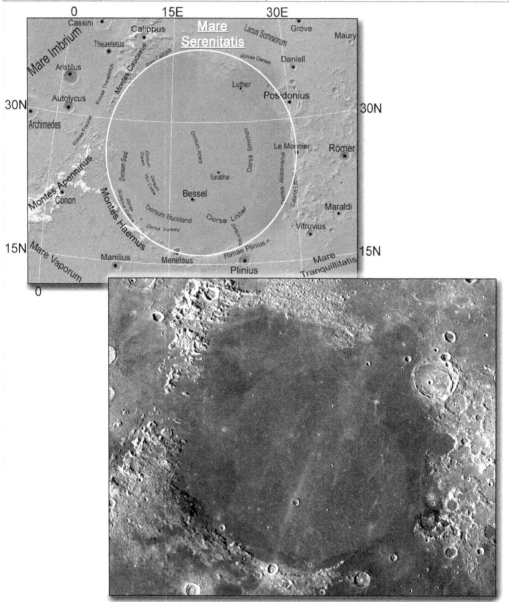

Maria

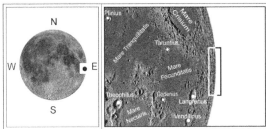

Mare Smythii	
Lat 1.71S	Long 87.05E
Diameter	373.97Km
Notes:	

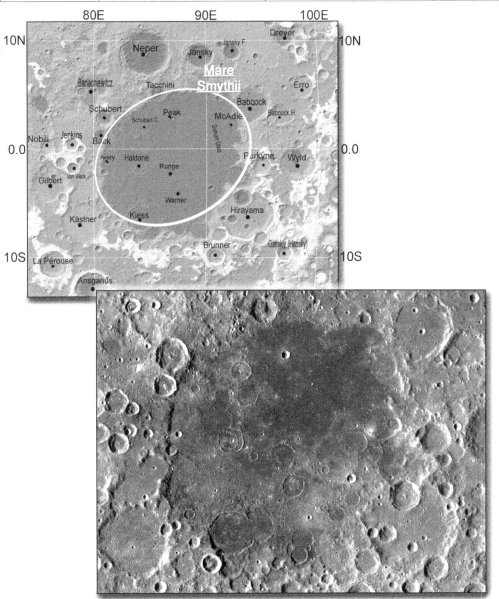

Maria

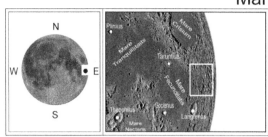

Mare Spumans	
Lat 1.3N	Long 65.3E
Diameter	143.13Km
Notes:	

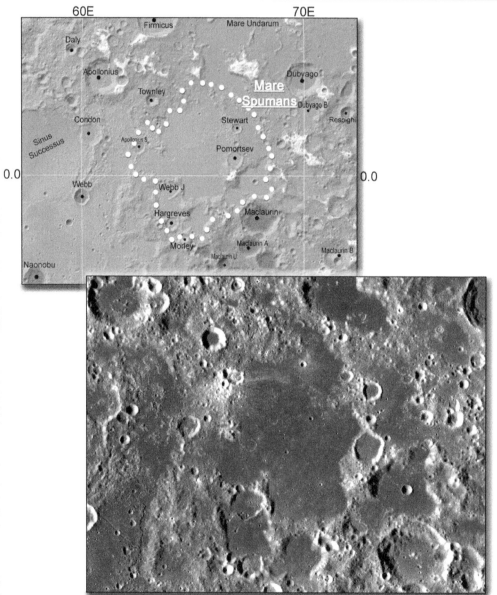

94

Maria

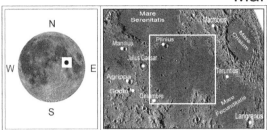

Mare Tranquillitatis	
Lat 8.35N	Long 30.83E
Diameter	875.75Km
Notes:	

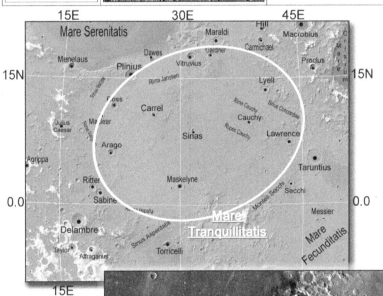

Maria

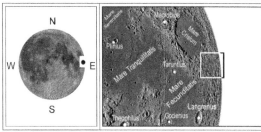

Mare Undarum	
Lat 7.49N	Long 68.66E
Diameter	244.84Km
Notes:	

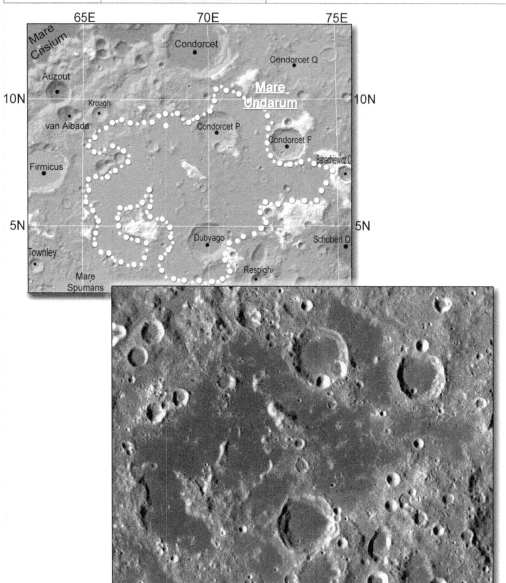

Maria

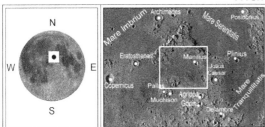

Mare Vaporum	
Lat 13.2N	Long 4.09E
Diameter	242.46Km
Notes:	

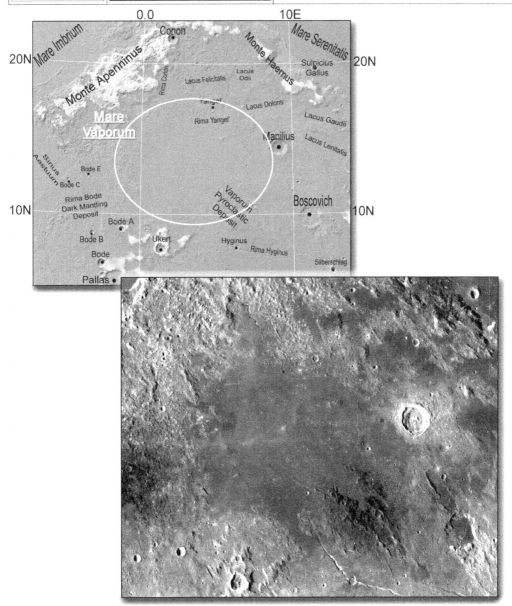

Mons

Mons

Mons translates from the Latin word for 'mountain'. In the lunar context the Mons can be seen as small mountain outcrops 'connected' to a much larger mountain range (see how Mons Ampère, for example, connects to Montes Apenninus), while in other instances they may be isolated (as seen with Mons Delisle). The highest mountain on the moon's Nearside is belie-ved to be Mons Huygens - coming in at some 5.5 km high (Everest on Earth is some 8.8 km high).

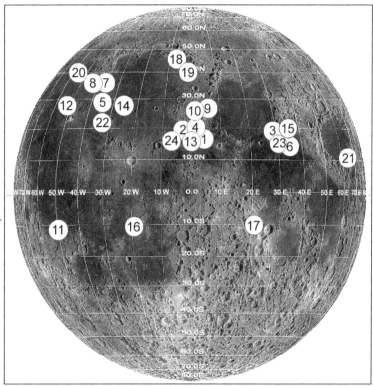

No.	Feature	No.	Feature	No.	Feature
1	Agnes	9	Hadley	17	Penck
2	Ampère	10	Hadley Delta	18	Pico
3	Argaeus	11	Hansteen	19	Piton
4	Bradley	12	Herodotus	20	Rümker
5	Delisle	13	Huygens	21	Usov
6	Esam	14	La Hire	22	Vinogradov
7	Gruithuisen Delta	15	Maraldi	23	Vitruvius
8	Gruithuisen Gamma	16	Moro	24	Wolff

Note: The above list of Mons represents those given in the International Astronomical Union list, but there are plenty of more mountains on the Moon that may in the future be given names and official designations.

Mons

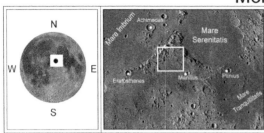

Mons Agnes	
Lat 18.66N	Long 5.34E
Diameter	0.0Km
Notes:	

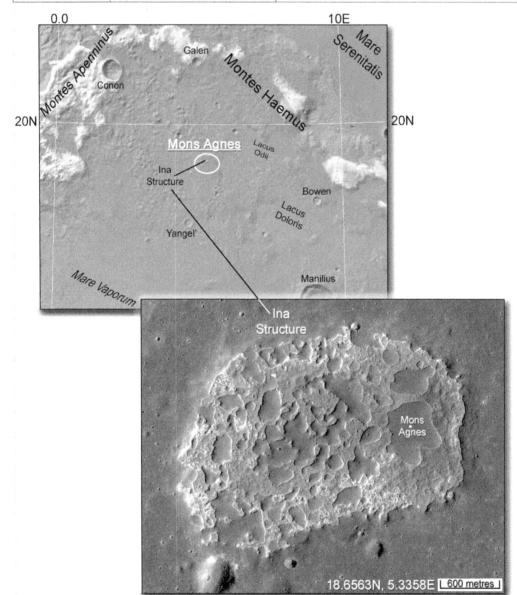

18.6563N, 5.3358E 600 metres

Mons

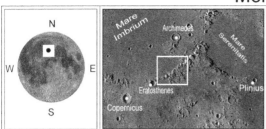

Mons Ampère	
Lat 19.32N	Long 3.71W
Diameter	29.96Km
Notes:	

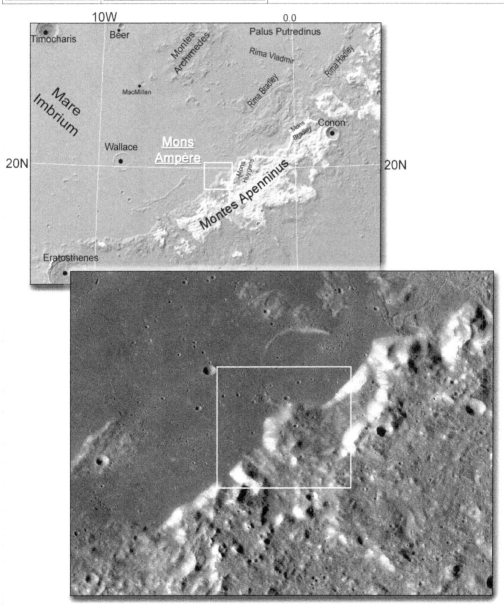

Mons

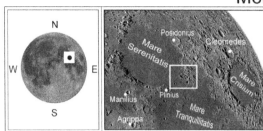

Mons Argaeus	
Lat 19.33N	Long 29.01E
Diameter	61.48Km
Notes:	

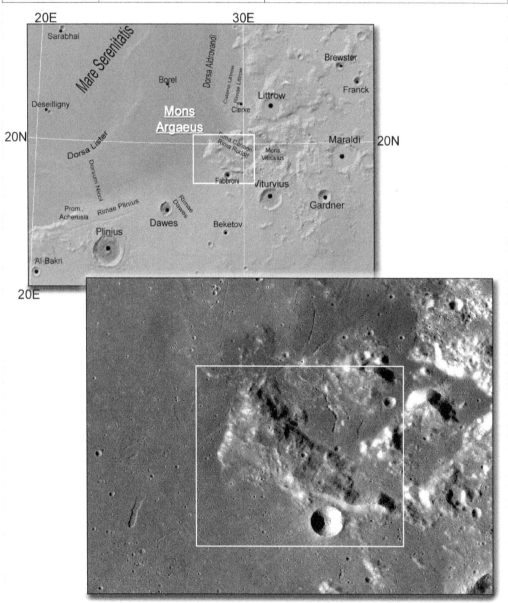

Mons

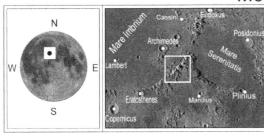

Mons Bradley	
Lat 21.73N	Long 0.38E
Diameter	76.49Km
Notes:	

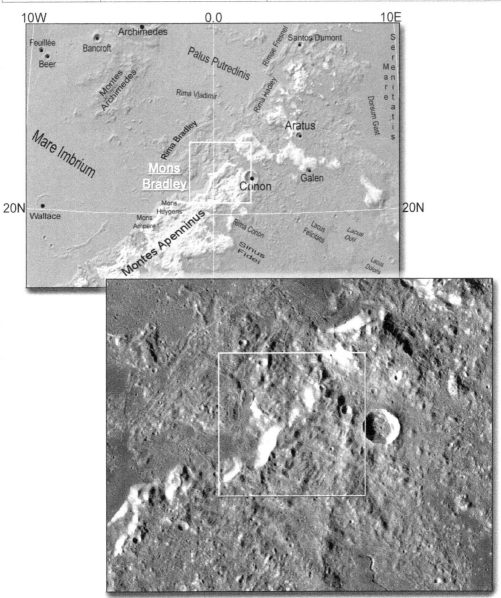

Mons

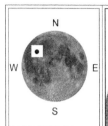
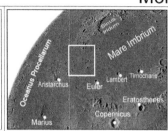

Mons Delisle	
Lat 29.42N	Long 35.79W
Diameter	32.42Km
Notes:	

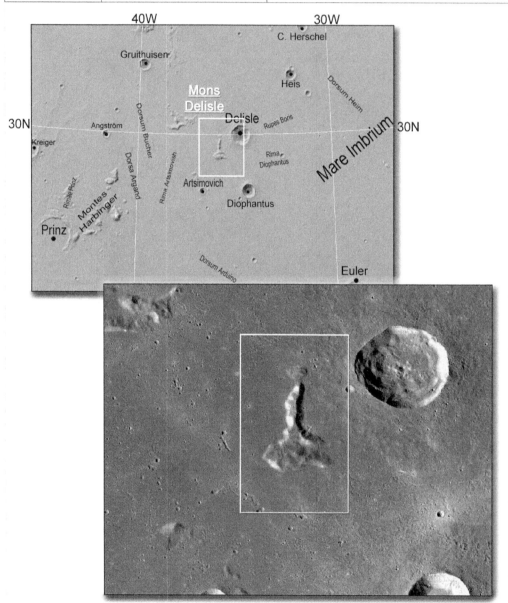

Mons

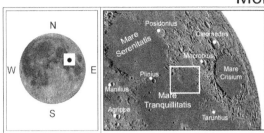

Mons Esam	
Lat 14.61N	Long 35.71E
Diameter	7.92Km
Notes:	

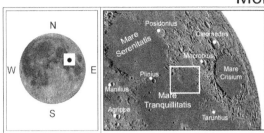

Mons

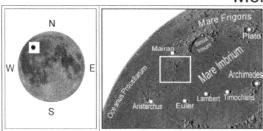

Mons Gruithuisen Delta	
Lat 36.07N	Long 39.59W
Diameter	27.24Km
Notes:	

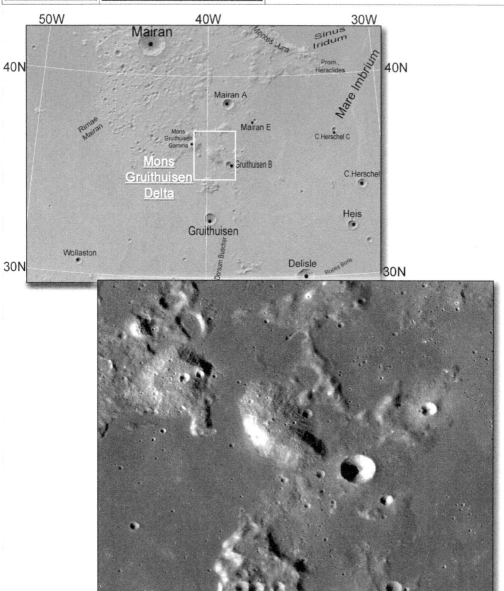

Mons

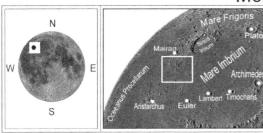

Mons Gruithuisen Gamma	
Lat 36.56N	Long 40.72W
Diameter	19.65Km
Notes:	

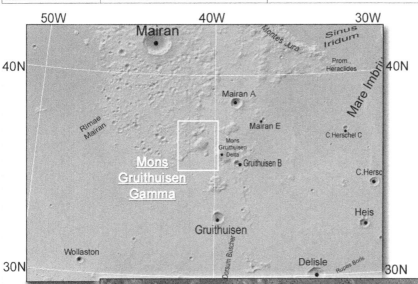

Mons

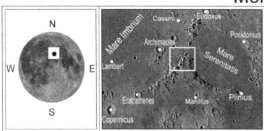

Mons Hadley	
Lat 26.69N	Long 4.12E
Diameter	26.4Km
Notes:	

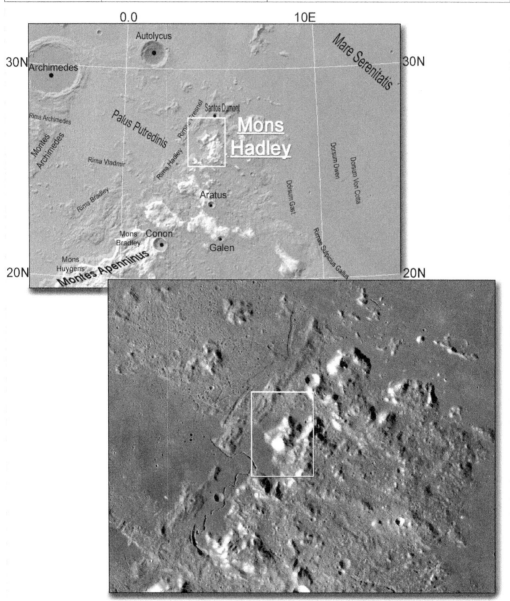

Mons

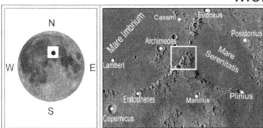

Mons Hadley Delta	
Lat 25.72N	Long 3.71E
Diameter	17.24Km
Notes:	

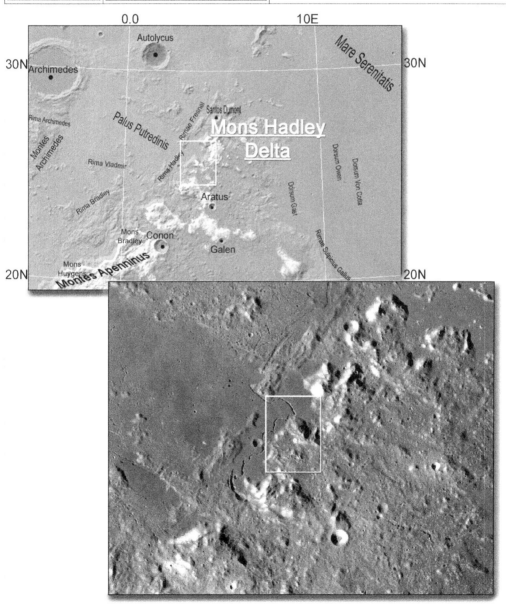

Mons

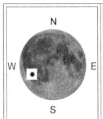

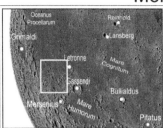

Mons Hansteen	
Lat 12.19S	Long 50.21W
Diameter	30.65Km
Notes:	

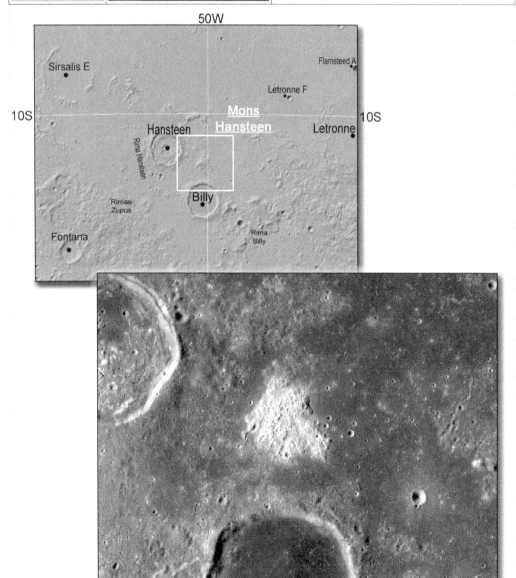

Mons

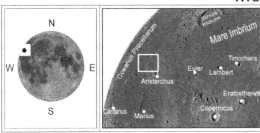

Mons Herodotus	
Lat 27.5N	Long 52.94W
Diameter	6.77Km
Notes:	

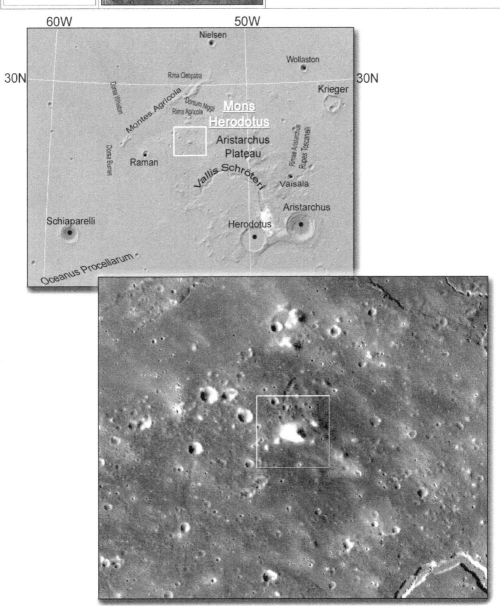

Mons

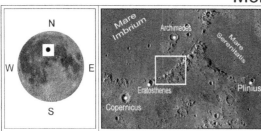

Mons Huygens	
Lat 19.92N	Long 2.86W
Diameter	41.97Km
Notes:	

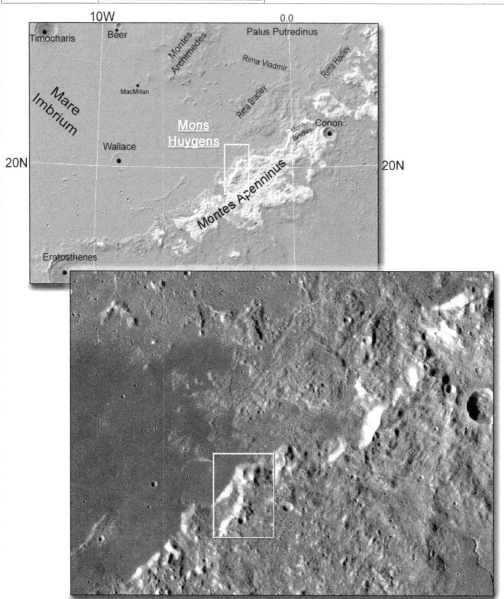

Mons

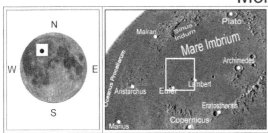

Mons La Hire	
Lat 27.66N	Long 25.51W
Diameter	21.71Km
Notes:	

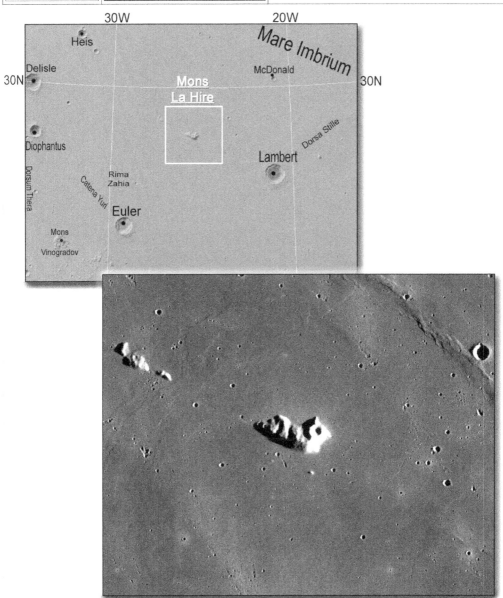

Mons

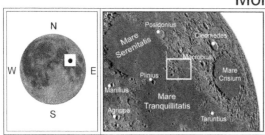

Mons Maraldi	
Lat 20.34N	Long 35.5E
Diameter	15.9Km
Notes:	

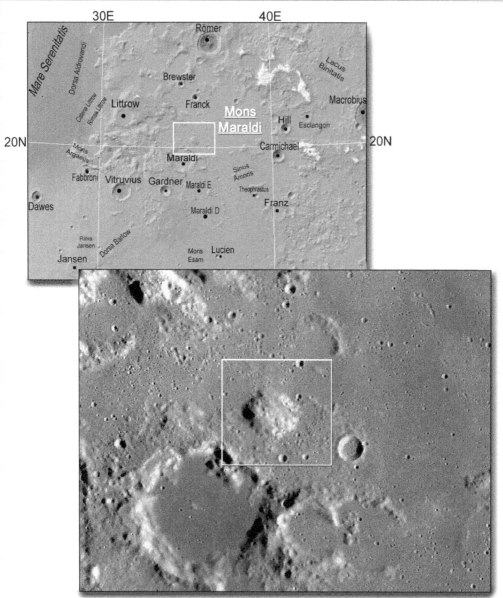

Mons

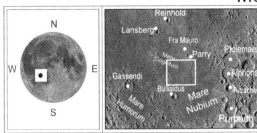

Mons Moro	
Lat 11.84S	Long 19.84W
Diameter	13.68Km
Notes:	

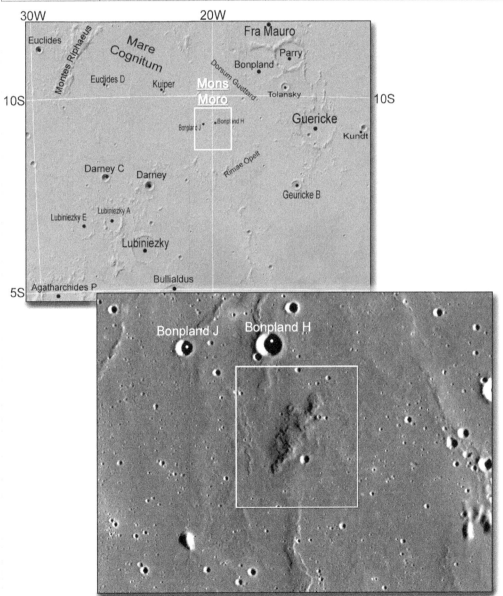

Mons

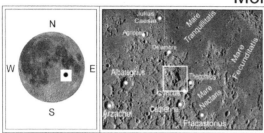

Mons Penck	
Lat 10.0S	Long 21.74E
Diameter	37.59Km
Notes:	

20E

Taylor
Alfraganus
Hypatia A
Sinus Asperitatis
Toricelli

Alfraganus C

Zöllner F

Zöllner

Mons Penck

Kant C

10S
10S

Kant
Mädler

Kant D
Ibn-Rushd
Theophilus

Descartes

Tacitus
Catharina
Mare Nectaris

Abulfeda

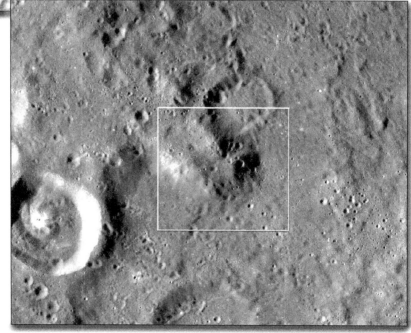

Mons

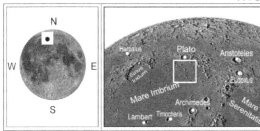

Mons Pico	
Lat 45.82N	Long 8.87W
Diameter	24.42Km
Notes:	

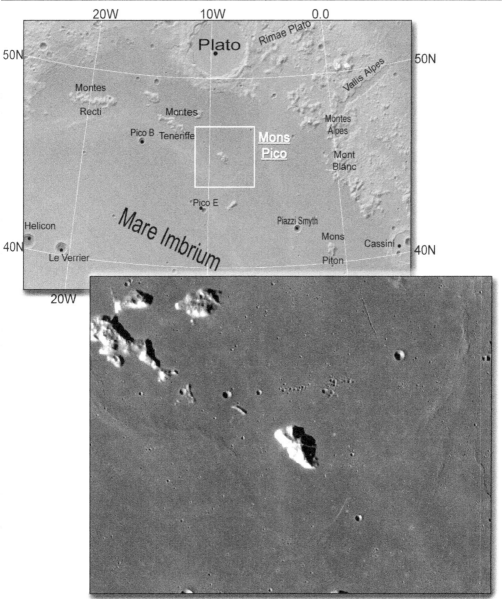

Mons

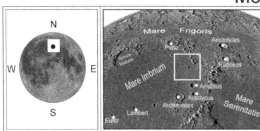

Mons Piton	
Lat 40.72N	Long 0.92W
Diameter	22.5Km
Notes:	

10W   0.0

Pico E

Piazzi Smyth

Prom. Deville

Prom. Agazziz

Cassini C

Cassini

40N   40N

Mare Imbrium

Kirch

Piton A

Piton B

Mons Piton

Theaetetus

Montes Spitzbergen

Aristillus

Mons

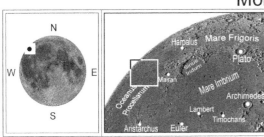

Mons Rümker	
Lat 40.76N	Long 58.38W
Diameter	73.25Km
Notes:	

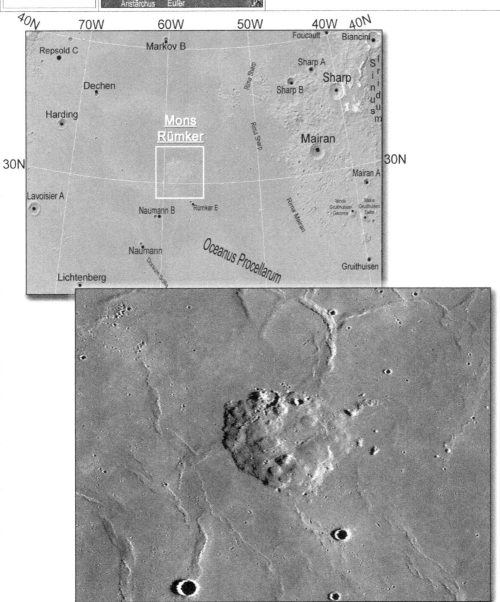

Mons

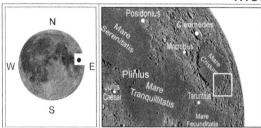

Mons Usov	
Lat 11.91N	Long 63.26E
Diameter	13.23Km
Notes:	

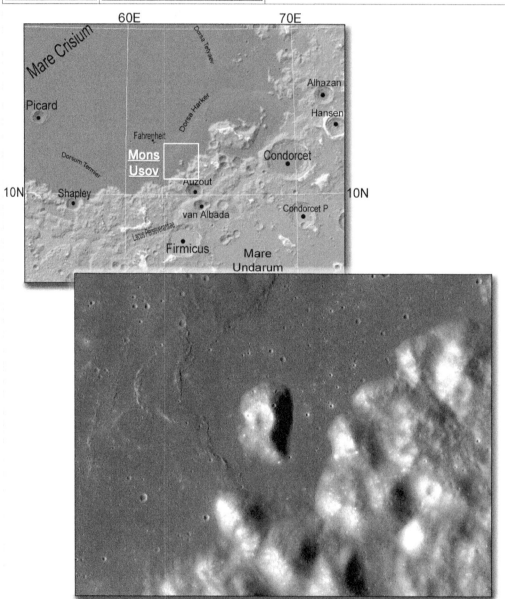

Mons

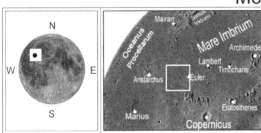

Mons Vinogradov	
Lat 22.35N	Long 32.52W
Diameter	28.73Km
Notes:	

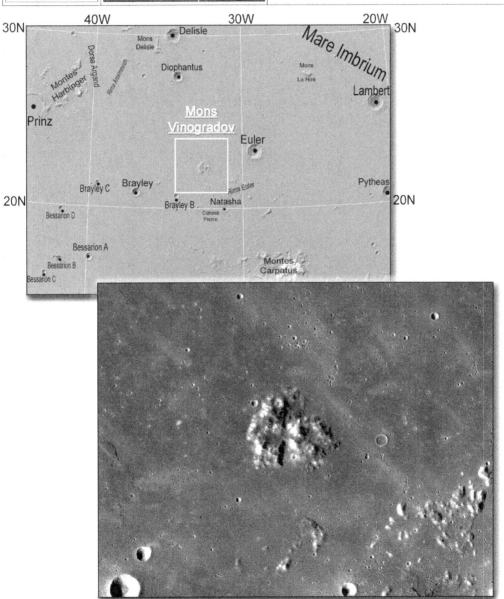

Mons

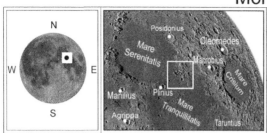

Mons Vitruvius	
Lat 19.33N	Long 30.74E
Diameter	44.28Km
Notes:	

Mons

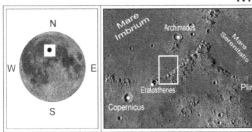

Mons Wolff	
Lat 16.88N	Long 6.8W
Diameter	32.87Km
Notes:	

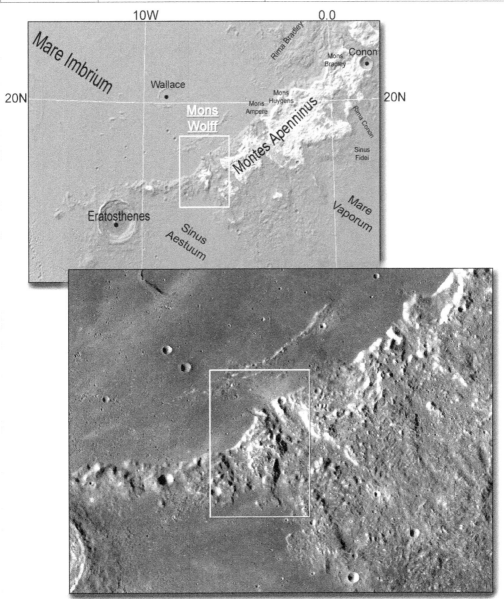

Mont

Mont

Mont translates from the French word for 'mount'. In the lunar context, Mont is similar to the Latin form of Mons (see Mons pages) as used for naming mountains on the Moon, and so no difference between the two words used is directly related to different features. The word Mont, as applied to Mont Blanc, literally translates to 'White Mountain'.

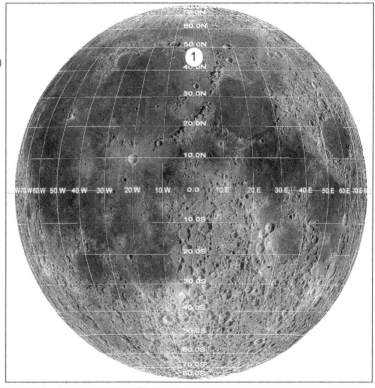

No.	Feature	No.	Feature	No.	Feature
1	Blanc				

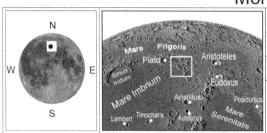

Mont Blanc	
Lat 45.41N	Long 0.44E
Diameter	21.57Km
Notes:	

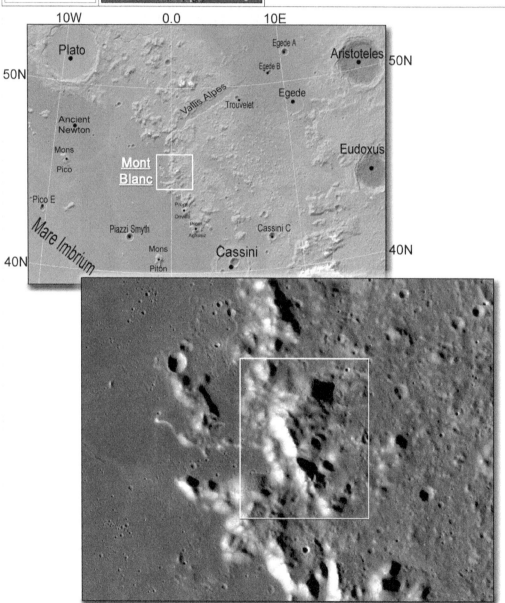

Montes

Montes

Montes is the plural form of the word 'Mons', which in Latin means 'mountain'. Most of the Montes make up mountain ranges that formed due to some of the major basins - essentially, they are their resultant rims (see, for example, Montes Apenninus), while others may be the ejecta from these basins flung far and wide (as believed with the Montes Taurus). Of course, some also are isolated (as seen with Montes Recti), which in some cases may just be related to inner basin rings that formed during the dynamics that produced them.

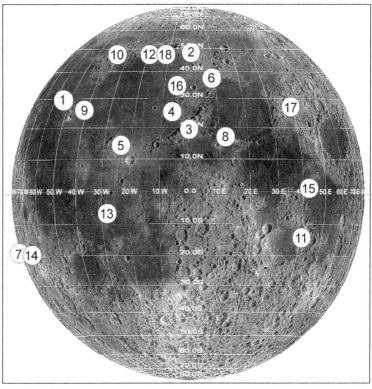

No.	Feature	No.	Feature	No.	Feature
1	Agricola	7	Cordillera	13	Riphaeus
2	Alpes	8	Haemus	14	Rook (inner & outer)
3	Apenninus	9	Harbinger	15	Seechi
4	Archimedes	10	Jura	16	Spitzbergen
5	Carpatus	11	Pyrenaeus	17	Taurus
6	Caucasus	12	Recti	18	Teneriffe

Note: The above list of Montes represents those given in the International Astronomical Union list, but there are plenty of more mountains on the Moon that may in the future be given names and official designations.

Montes

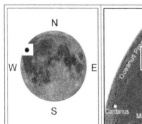

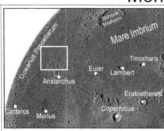

Montes Agricola	
Lat 29.06N Long 54.07W	
Diameter	159.76Km
Notes:	

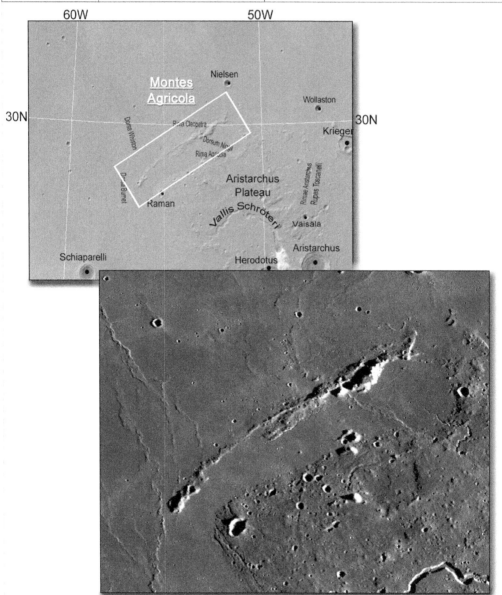

128

Montes

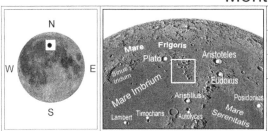

Montes Alpes	
Lat 48.36N	Long 0.58W
Diameter	334.48Km
Notes:	

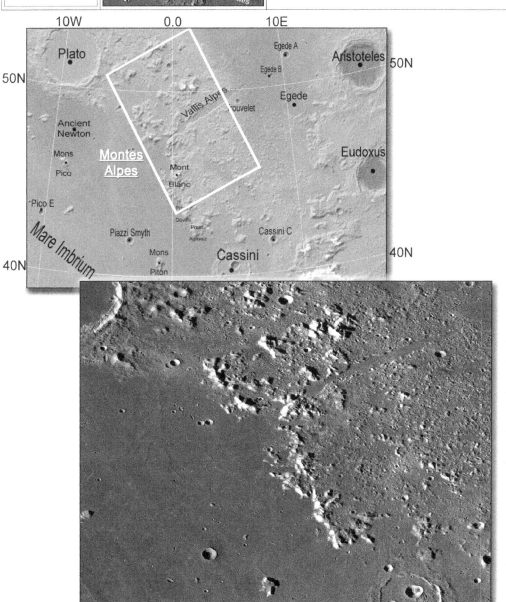

Montes

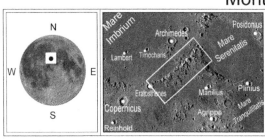

Montes Apenninus	
Lat 19.87N	Long 0.03E
Diameter	599.67Km
Notes:	

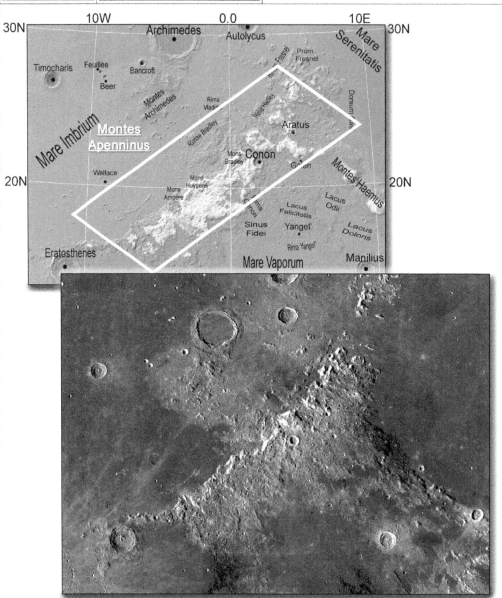

Montes

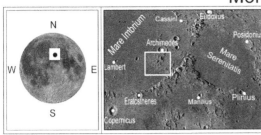

Montes Archimedes	
Lat 25.39N	Long 5.25W
Diameter	146.54Km
Notes:	

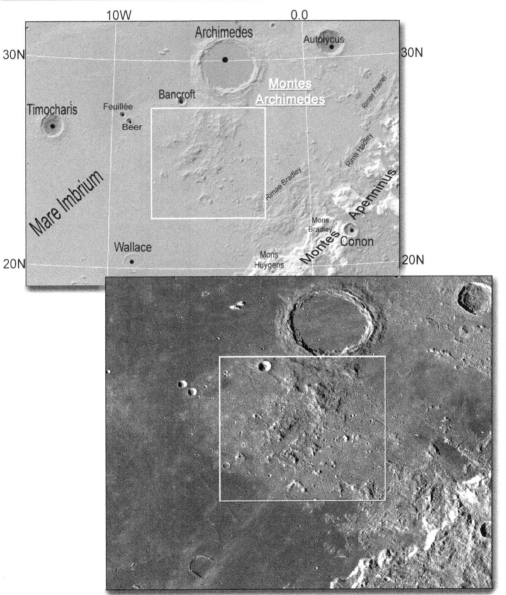

Montes

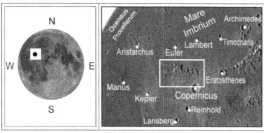

Montes Carpatus	
Lat 14.57N	Long 23.62W
Diameter	333.59Km
Notes:	

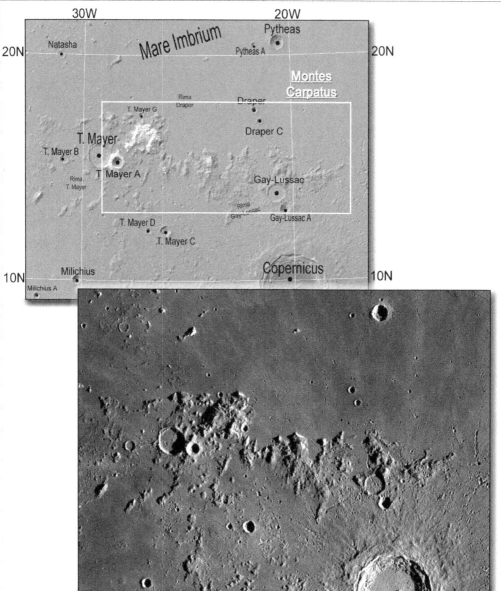

132

Montes

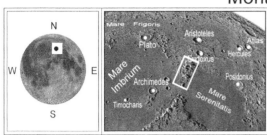

Montes Caucasus	
Lat 37.52N	Long 9.93E
Diameter	443.51Km
Notes:	

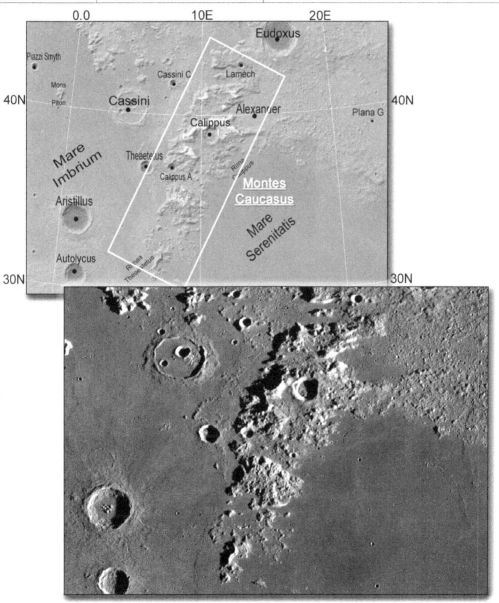

133

Montes

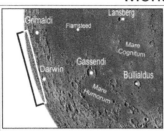

Montes Cordillera	
Lat 19.44S	Long 94.93W
Diameter	963.5Km
Notes:	

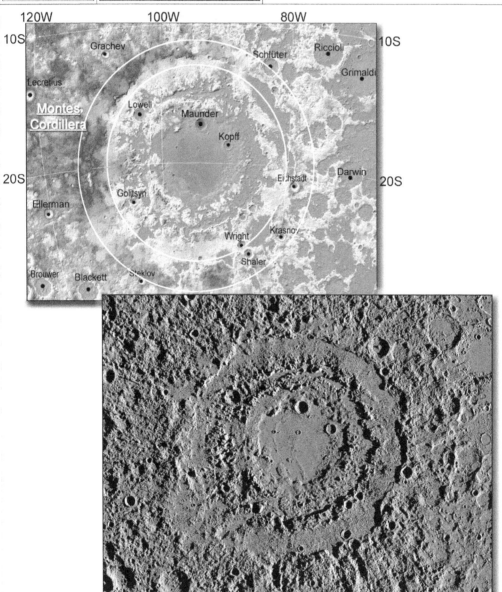

Montes

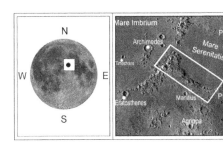

Montes Haemus	
Lat 17.11N	Long 12.03E
Diameter	384.66Km
Notes:	

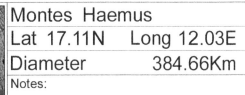

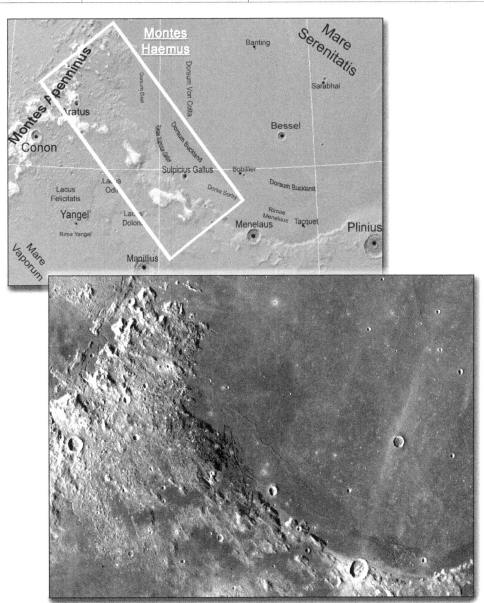

Montes

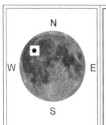

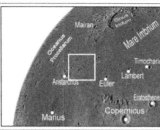

Montes Harbinger	
Lat 26.89N	Long 41.29W
Diameter	92.7Km
Notes:	

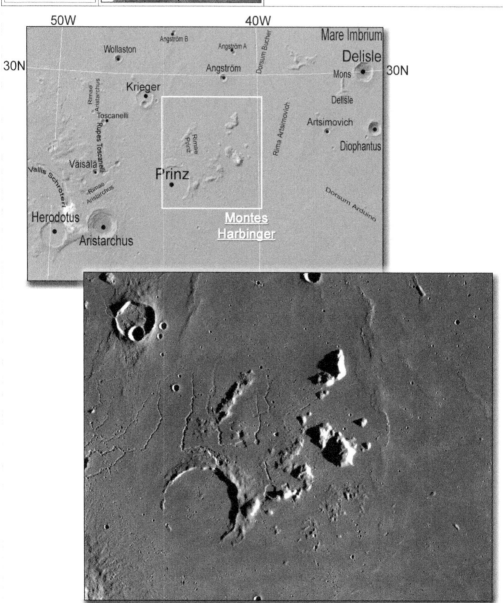

Montes

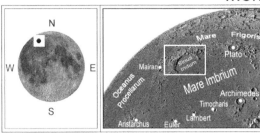

Montes Jura	
Lat 47.49N	Long 36.11W
Diameter	420.8Km
Notes:	

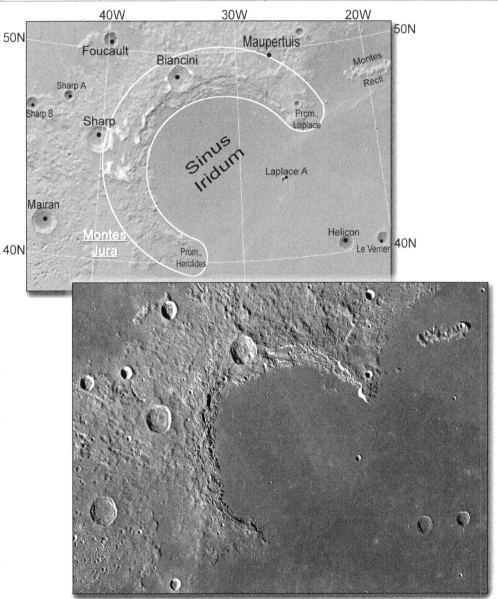

Montes

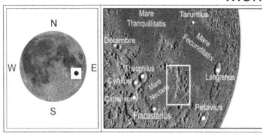

Montes Pyrenaeus	
Lat 14.05S	Long 41.51E
Diameter	251.33Km
Notes:	

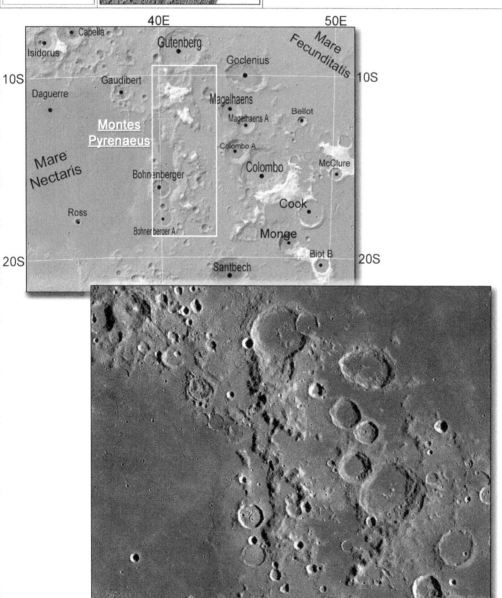

Montes

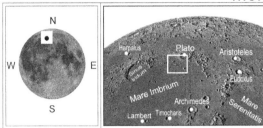

Montes Recti	
Lat 48.3N	Long 19.72W
Diameter	83.24Km
Notes:	

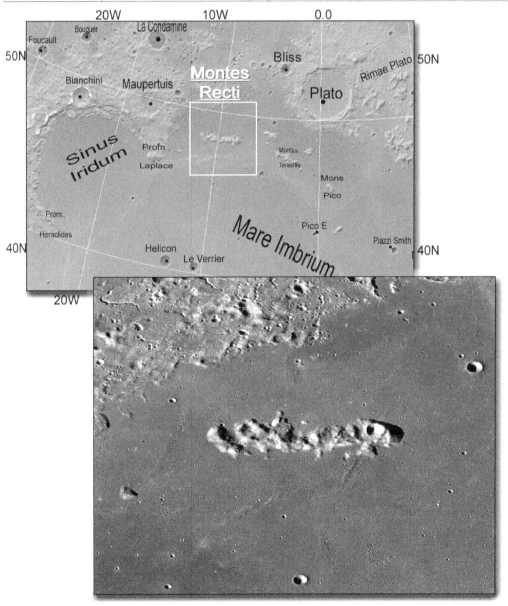

Montes

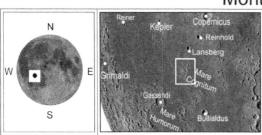

Montes Riphaeus	
Lat 7.48S	Long 27.6W
Diameter	190.12Km
Notes:	

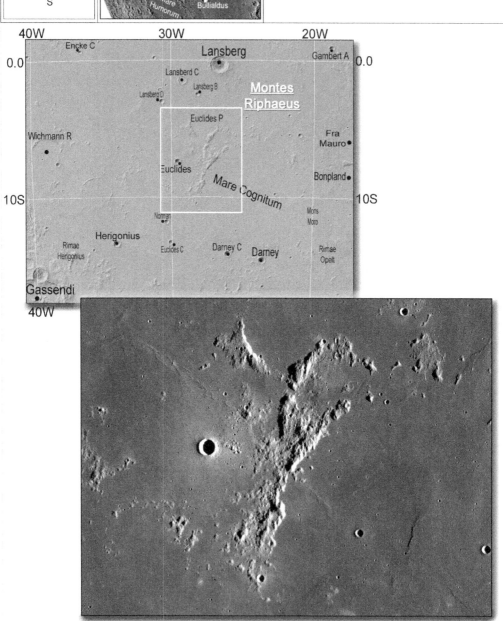

Montes

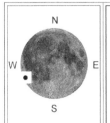

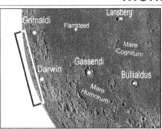

Montes Rook (inner)	
Lat 19.49S	Long 94.95W
Diameter	682.28Km
Notes:	

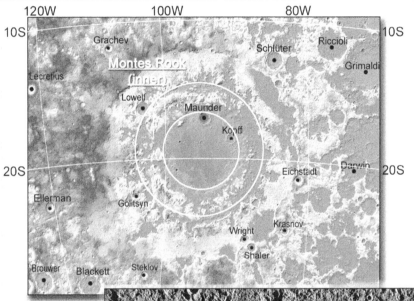

Montes

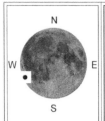

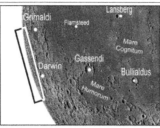

Montes Rook (outer)	
Lat 19.49S	Long 94.95W
Diameter	682.28Km
Notes:	

142

Montes

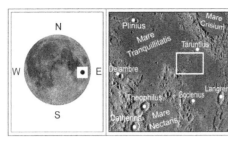

Montes Secchi	
Lat 2.72N	Long 43.17E
Diameter	52.47Km
Notes:	

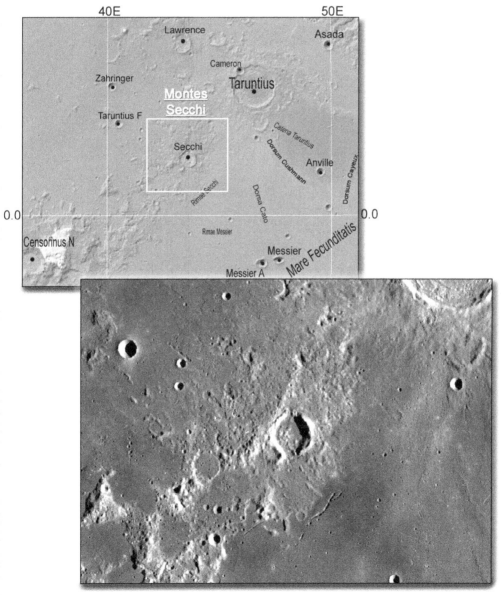

Montes

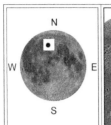
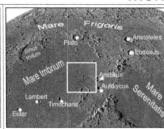

Montes Spitzbergen	
Lat 34.47N	Long 5.21W
Diameter	59.19Km
Notes:	

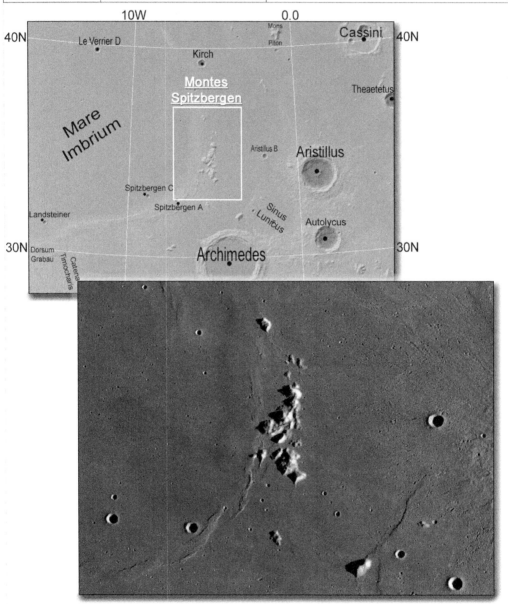

Montes

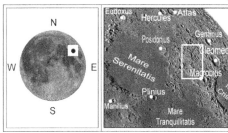

Montes Taurus	
Lat 27.32N	Long 40.34E
Diameter	166.16Km
Notes:	

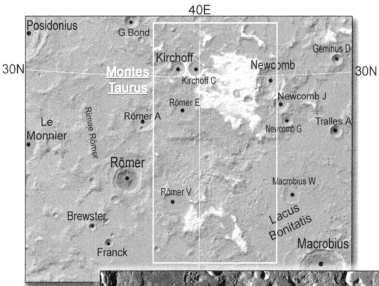

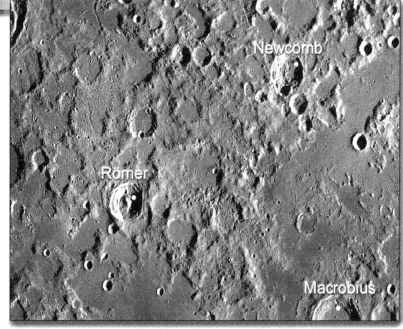

Montes

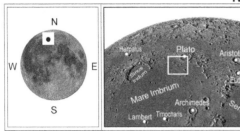

Montes Teneriffe	
Lat 47.89N	Long 13.19W
Diameter	111.98Km
Notes:	

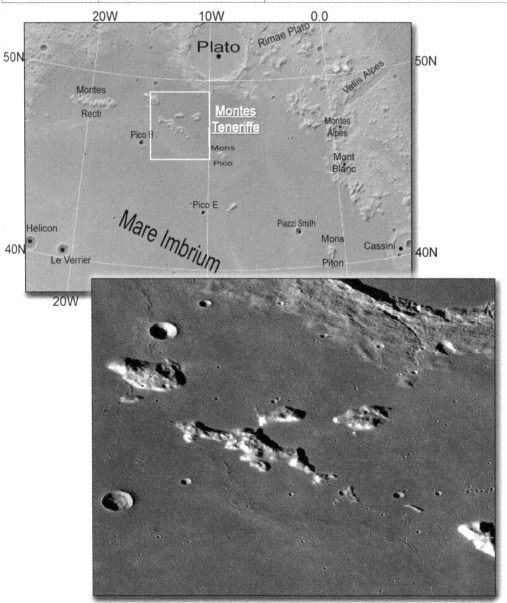

Palus

Palus

Palus translates from the Latin word for 'marsh' or 'swamp'. In the lunar context the Palus seem to differ no more to the Lacus (see Lacus Pages), which appear as isolated patches of mare material usually 'connected' to larger mares (Palus Somni being somewhat of an exception whose texture and colour does look, literally, like a marsh).

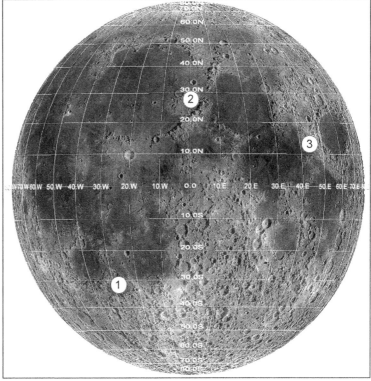

No.	Feature			No.	Feature	
1	Epidemiarum	-	Marsh of Epidemics	3	Somni	-	Marsh of Sleep
2	Putredinis	-	Marsh of Decay			

Note: The above list of Palus represents those given in the International Astronomical Union list, but there are plenty more on the Nearside's face that may in the future be given official designations.

Palus

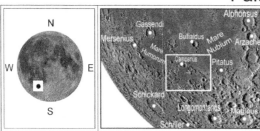

Palus Epidemiarum	
Lat 32.0S	Long 27.54W
Diameter	300.38Km
Notes:	

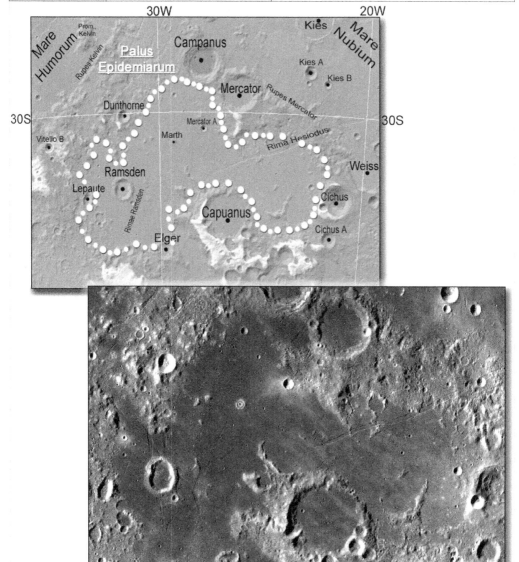

Palus

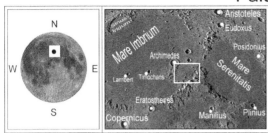

Palus Putredinis	
Lat 27.26N	Long 0.0
Diameter	180.45Km
Notes:	

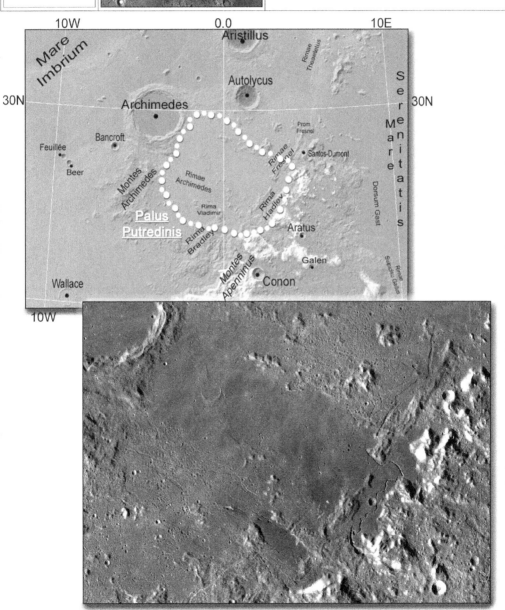

Palus

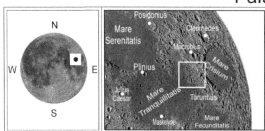

Palus Somni	
Lat 13.69N	Long 44.72E
Diameter	163.45Km
Notes:	

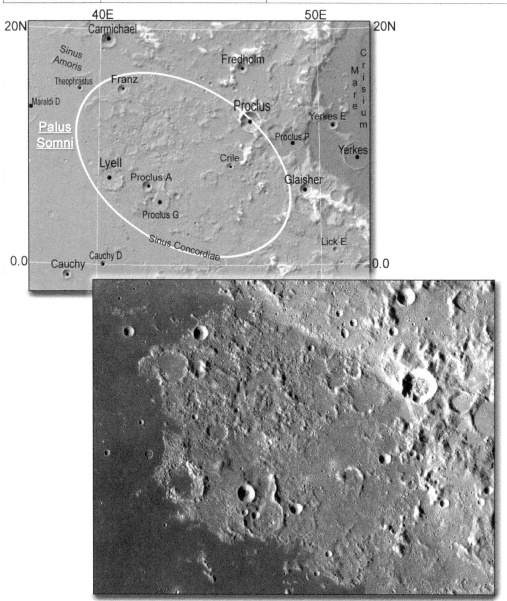

Promontorium

Promontorium

Promontorium translates from the Latin to the word 'cape'. In the lunar context it applies to headlands connected mostly to major mountain ranges on the Moon - made from rims, ejecta, or from the local pre-existant highlands' material created early during the moon's form- ation.

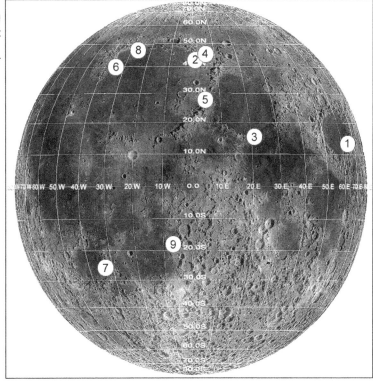

No.	Feature	No.	Feature	No.	Feature
1	Agarum	4	Deville	7	Kelvin
2	Agassiz	5	Fresnel	8	Laplace
3	Archerusia	6	Heraclides	9	Taenarium

Note: The above list of Promontorium represents those given in the International Astronomical Union list, but there are plenty more on the Nearside's face that may in the future be given official designations.

151

Promontorium

Promontorium Agarum	
Lat 13.87N	Long 65.73E
Diameter	62.46Km
Notes:	

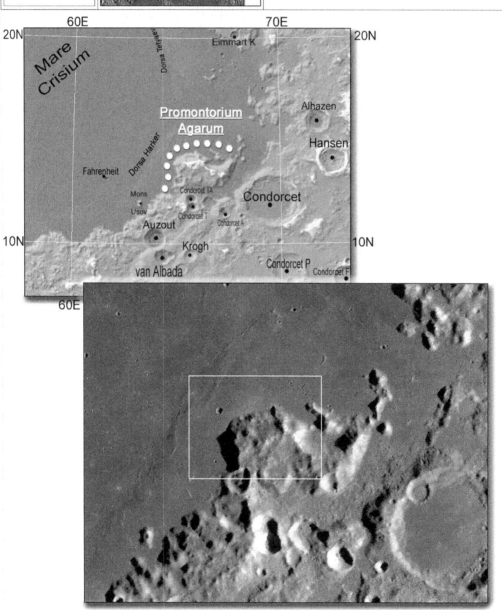

Promontorium

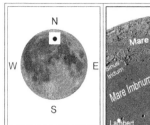

Promontorium Agassiz	
Lat 42.4N	Long 1.77E
Diameter	18.84Km
Notes:	

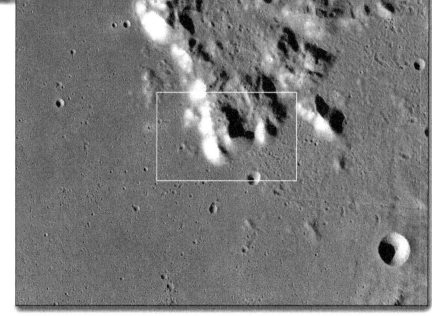

Promontorium

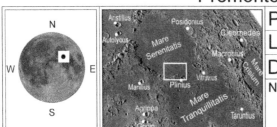

Promontorium Archerusia	
Lat 16.8N	Long 21.94E
Diameter	11.21Km
Notes:	

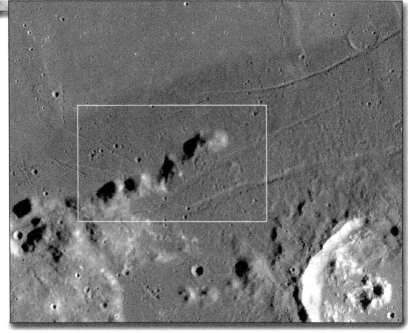

Promontorium

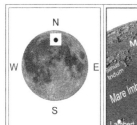

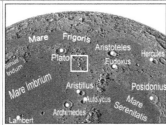

Promontorium Deville	
Lat 43.31N	Long 1.14E
Diameter	16.56Km
Notes:	

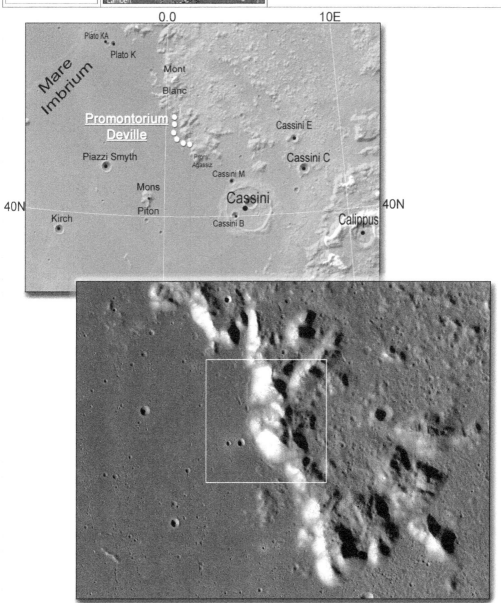

Promontorium

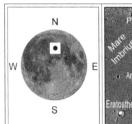

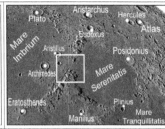

Promontorium Fresnel	
Lat 28.63N	Long 4.75E
Diameter	20.0Km
Notes:	

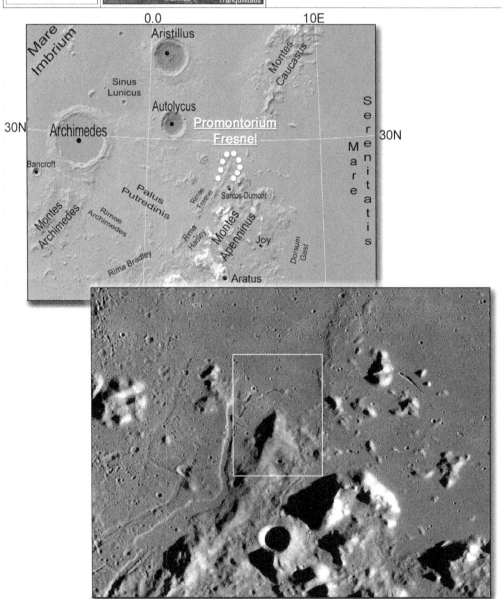

Promontorium

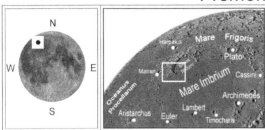

Promontorium Heraclides	
Lat 40.6N	Long 34.1W
Diameter	50.0Km
Notes:	

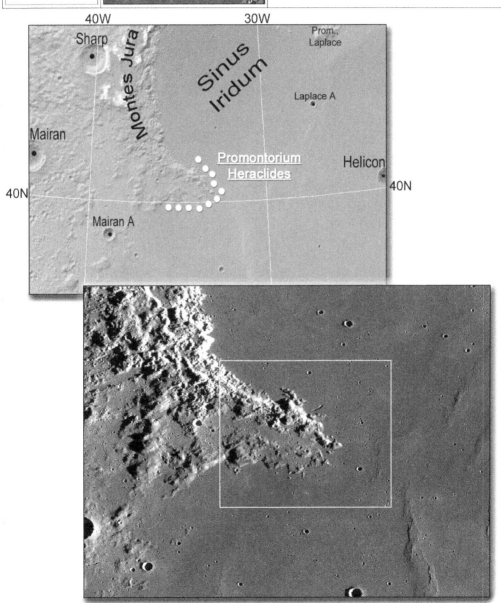

Promontorium

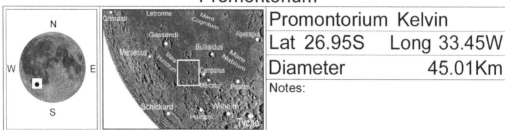

Promontorium Kelvin	
Lat 26.95S	**Long 33.45W**
Diameter	**45.01Km**
Notes:	

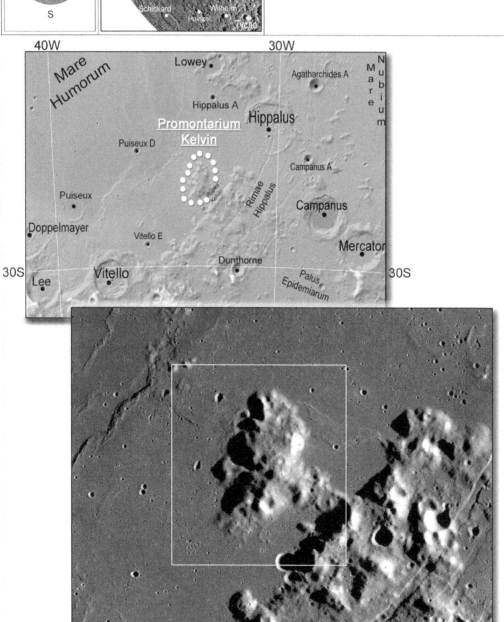

Promontorium

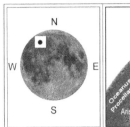

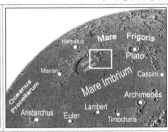

Promontorium Laplace	
Lat 46.84N	Long 25.51W
Diameter	50.0Km
Notes:	

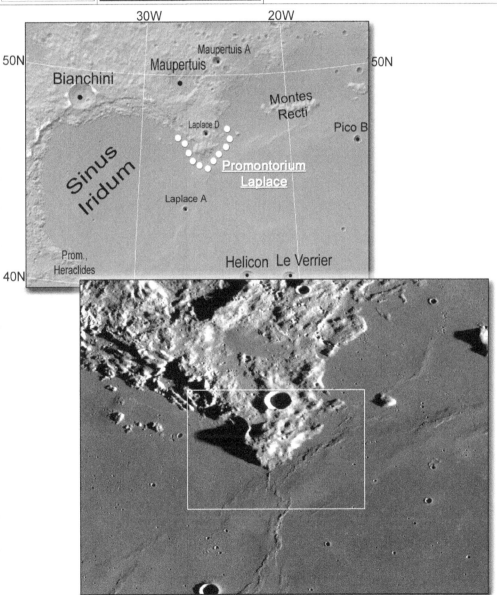

Promontorium

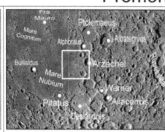

Promontorium Taenarium	
Lat 18.63S	Long 7.34W
Diameter	70.0Km
Notes:	

10W   0.0

Davy • Davy A   Klein •

Alphonsus
• Rimae
Alphonsus

Guericke B
•

Lassel
•   Alpetragius
                          •

Mare
Nubium   Promontorium
                Taenarium   Arzachel
                                         • Rimae
                                           Arzachel   Parrot C
                                                              •

20S   20S

Nicollet   Delauney
   •   Thebit A
                          • Thebit
         Rima Birt Rupes Recta
                    Birt   La Caille
                     •   •
              Thebit P

160

Rima

Rima

Rima on the lunar surface usually refers to just a singular rille (a volcanic feature in which lavas rose onto the moon's surface through channels - see the Rimae page description for more). Rima appear linear-like at times e.g. Rima Ariadaeus, while others can tend towards bending e.g. Rima Carmen to sinuous-like e.g. Rima Marian.

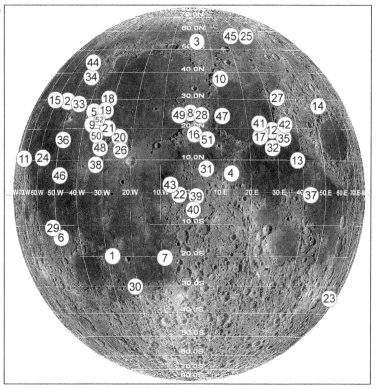

No.	Feature	No.	Feature	No.	Feature
1	Agatharchides	19	Diophantus	37	Messier
2	Agricola	20	Draper	38	Milichius
3	Archytas	21	Euler	39	Oppolzer
4	Ariadaeus	22	Flammarion	40	Réaumur
5	Artsimovich	23	Furnerius	41	Reiko
6	Billy	24	Galilaei	42	Rudolf
7	Birt	25	Gärtner	43	Schröter
8	Bradley	26	Gay-Lussac	44	Sharp
9	Brayley	27	G. Bond	45	Sheepshanks
10	Calippus	28	Hadley	46	Suess
11	Cardanus	29	Hansteen	47	Sung-Mei
12	Carmen	30	Hesiodus	48	T. Mayer
13	Cauchy	31	Hyginus	49	Vladimir
14	Cleomedes	32	Jansen	50	Wan-Yu
15	Cleopatra	33	Krieger	51	Yangel'
16	Conon	34	Mairan	52	Zahia
17	Dawes	35	Marcello		
18	Delisle	36	Marius		

Note: The above list of Rima represents those given in the International Astronomical Union list, but there are plenty more on the Nearside's face that may in the future be given official designations.

Rima

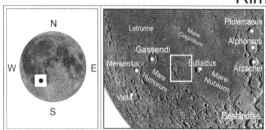

Rima Agatharchides	
Lat 20.38S	Long 28.56W
Diameter	54.25Km
Notes:	

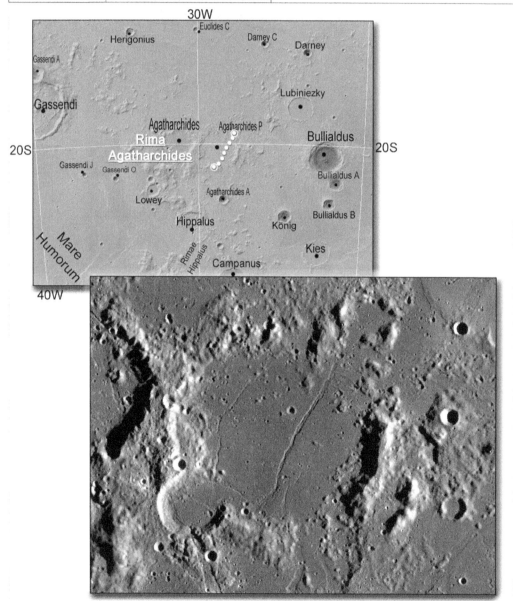

Rima

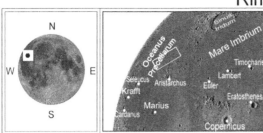

Rima Agricola	
Lat 29.25N Long 53.42W	
Diameter	125.08Km
Notes:	

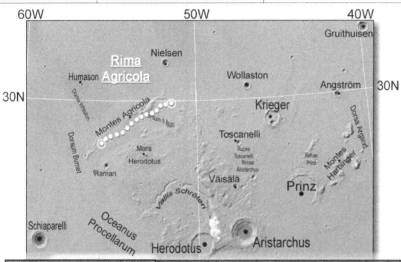

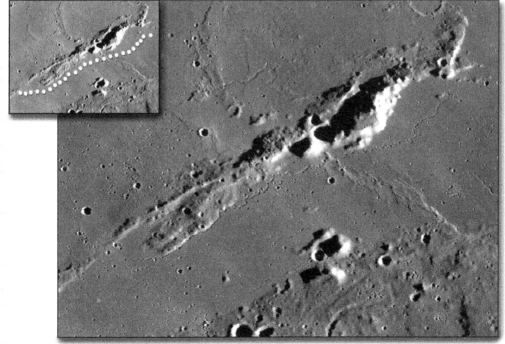

Rima

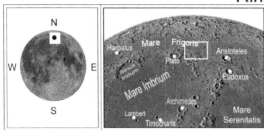

Rima Archytas	
Lat 53.63N	Long 3.0E
Diameter	90.18Km
Notes:	

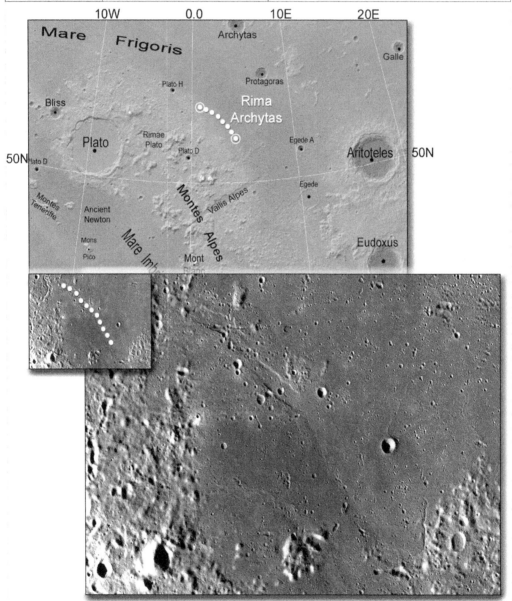

Rima

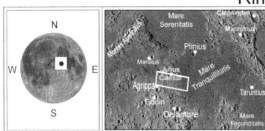

Rima Ariadaeus	
Lat 6.48N	Long 13.44E
Diameter	247.45Km
Notes:	

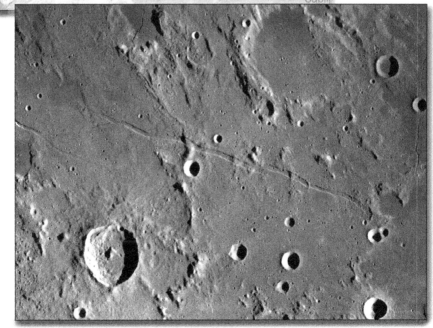

Rima

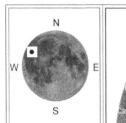

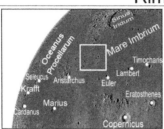

Rima Artsimovich	
Lat 26.66N	Long 38.65W
Diameter	68.06Km
Notes:	

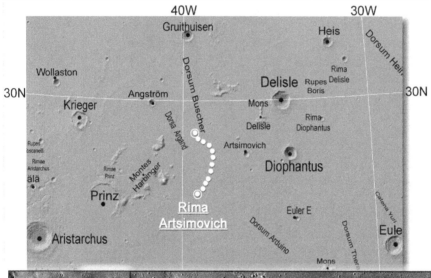

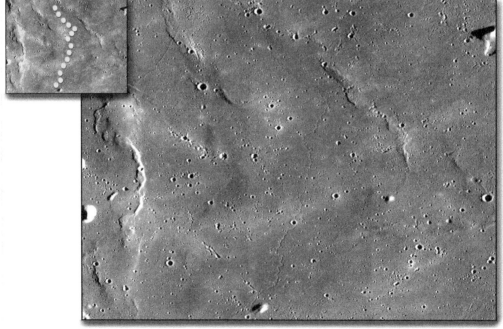

166

Rima

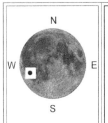

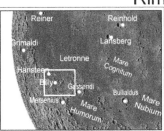

Rima Billy	
Lat 14.74S	Long 48.04W
Diameter	69.82Km
Notes:	

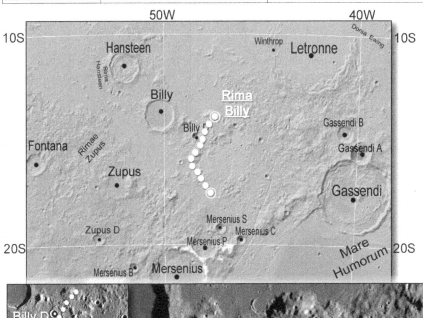

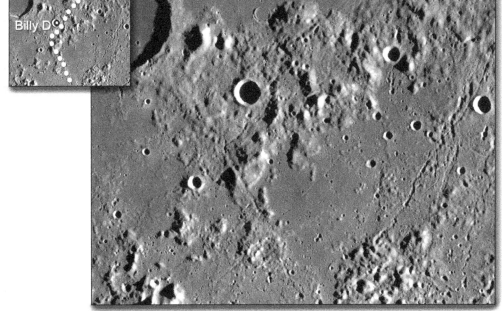

Rima

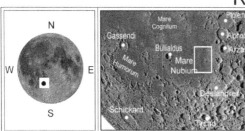

Rima Birt	
Lat 21.4S	Long 9.28W
Diameter	54.18Km
Notes:	

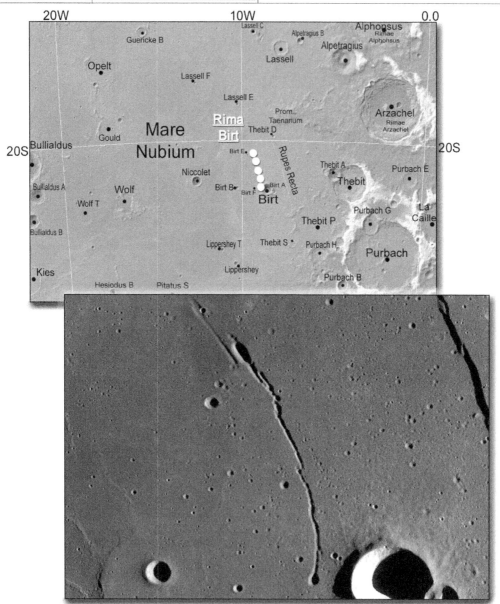

Rima

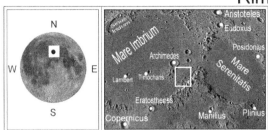

Rima Bradley	
Lat 24.17N	Long 0.6W
Diameter	133.76Km
Notes:	

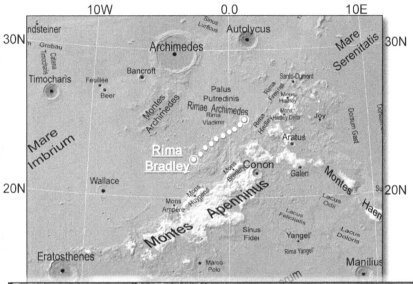

Rima

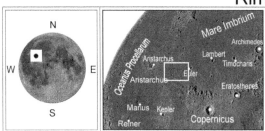

Rima Brayley	
Lat 22.3N	Long 36.35W
Diameter	327.26Km
Notes:	

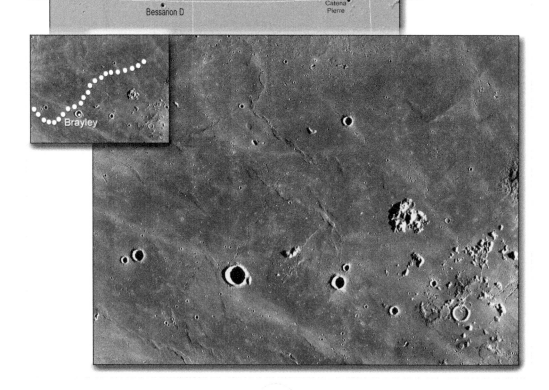

Rima

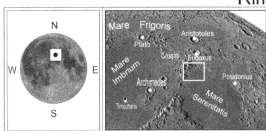

Rima Calippus	
Lat 37.03N	Long 12.66E
Diameter	40.0Km
Notes:	

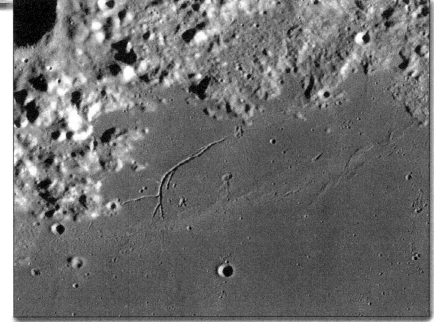

Rima

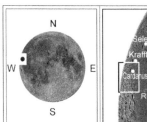

Rima Cardanus

Lat 11.32N	Long 71.14W
Diameter	221.93Km

Notes:

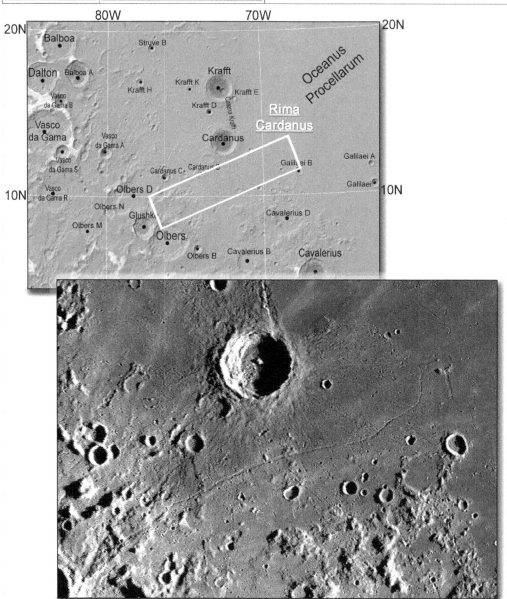

Rima

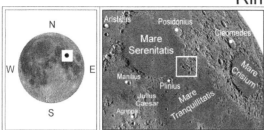

Rima Carmen	
Lat 19.95N Long 29.3E	
Diameter	15.02Km
Notes:	

173

Rima

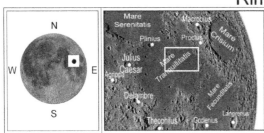

Rima Cauchy	
Lat 10.42N	Long 38.07E
Diameter	167.0Km
Notes:	

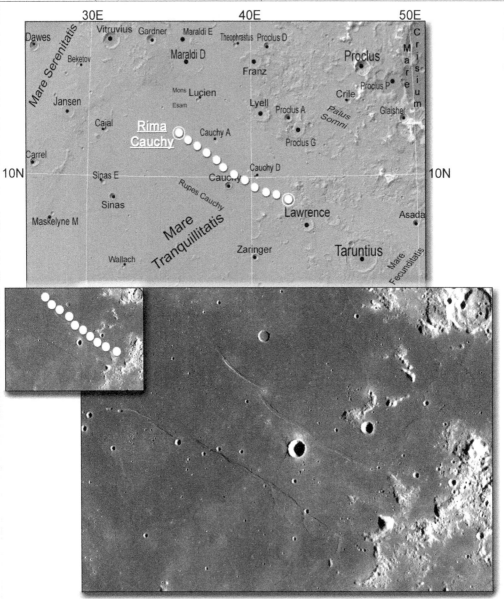

Rima

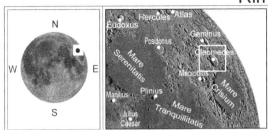

Rima Cleomedes	
Lat 27.98N	Long 56.51E
Diameter	45.54Km
Notes:	

50E   60E

Geminus

Geminus F

Geminus G

30N — Geminus D

Burckhardt

Burckhardt A   Cleomedes D   30N

Debes

Debes B

Tralles

Cleomedes A

Rima Cleomedes

Debes A   Cleomedes E

Tralles A

Tralles B

Cleomedes B

Cleomedes

Delmotte

Eimmart

Cleomedes G

Macrobius

Eimmart C

Mare Crisium

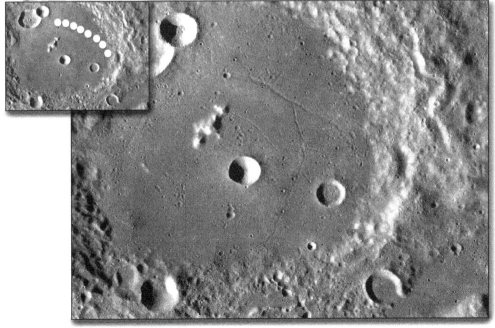

Rima

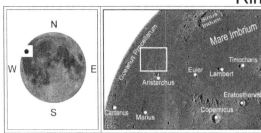

Rima Cleopatra	
Lat 30.03N	Long 53.8W
Diameter	14.66Km
Notes:	

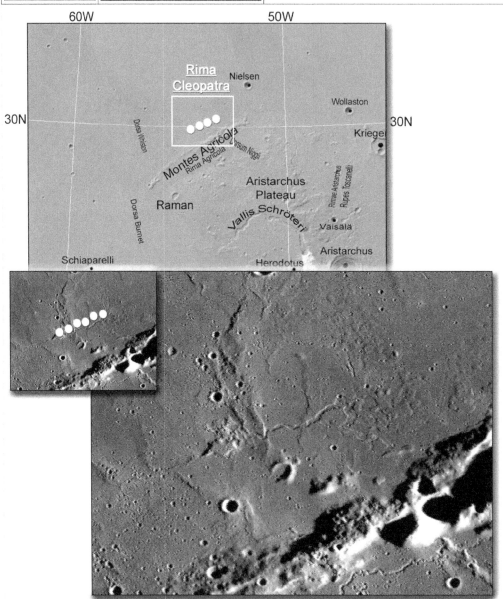

Rima

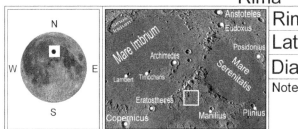

Rima Conon	
Lat 18.69N	Long 1.85E
Diameter	37.32Km
Notes:	

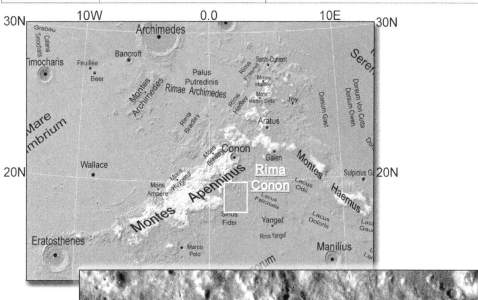

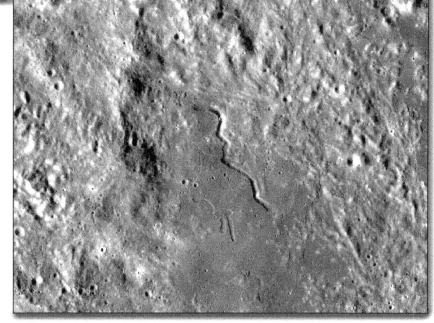

Rima

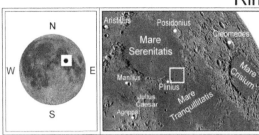

Rima Dawes	
Lat 17.58N	Long 26.63E
Diameter	15.0Km
Notes:	

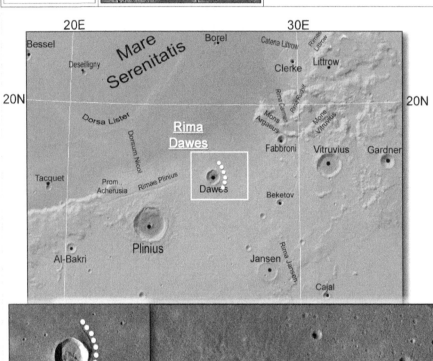

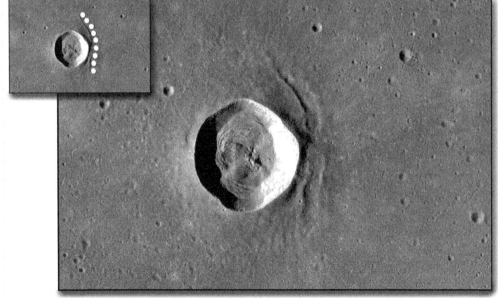

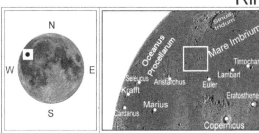

Rima Delisle	
Lat 30.87N	Long 32.35W
Diameter	57.6Km
Notes:	

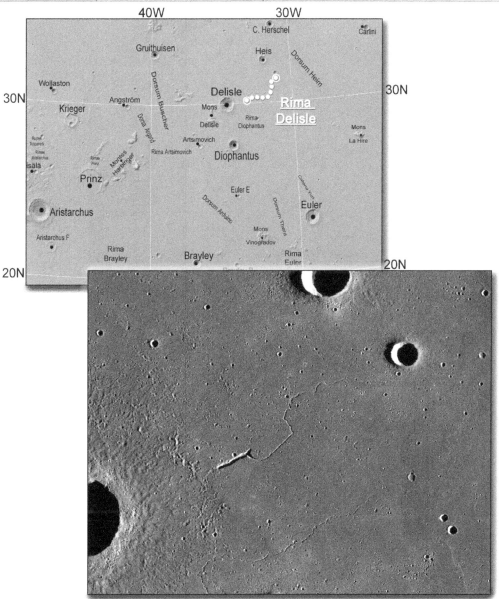

Rima

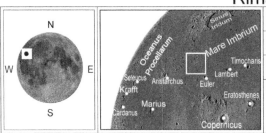

Rima Diophantus	
Lat 28.7N	Long 33.67W
Diameter	201.5Km
Notes:	

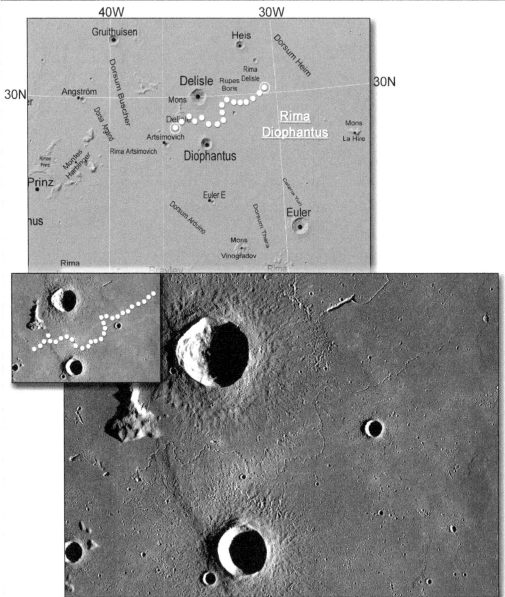

Rima

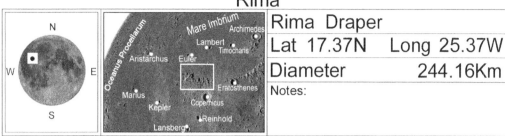

Rima Draper	
Lat 17.37N	Long 25.37W
Diameter	244.16Km
Notes:	

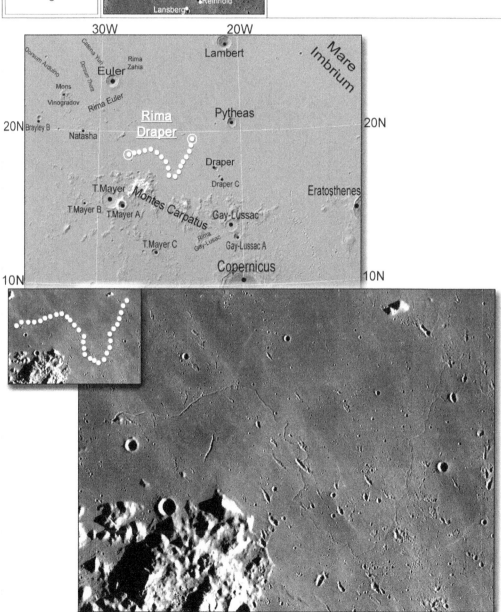

Rima

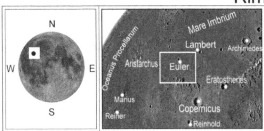

Rima Euler	
Lat 21.08N	Long 30.31W
Diameter	104.97Km
Notes:	

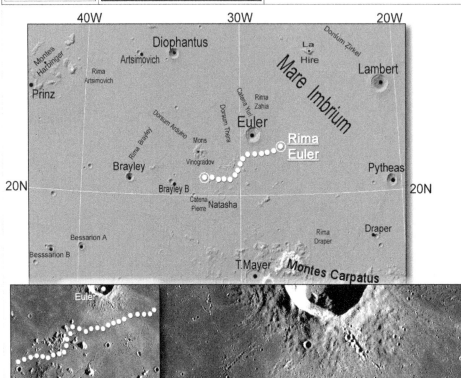

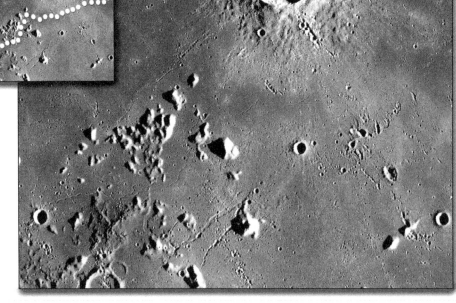

Rima

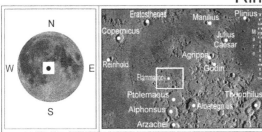

Rima Flammarion	
Lat 2.38S	Long 4.67W
Diameter	49.75Km
Notes:	

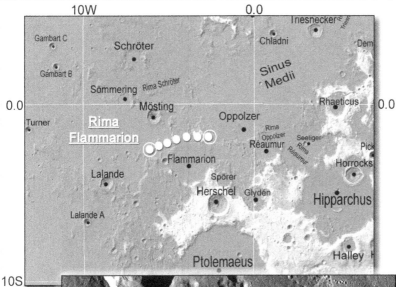

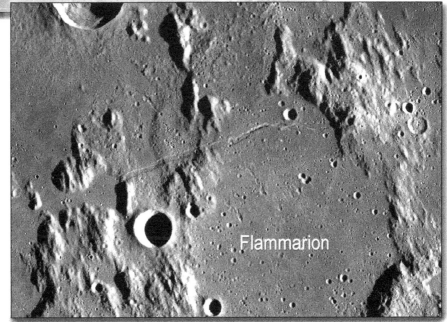

Rima

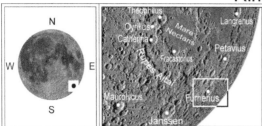

Rima Furnerius	
Lat 35.3S	Long 61.17E
Diameter	65.85Km
Notes:	

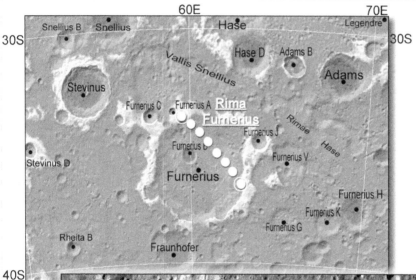

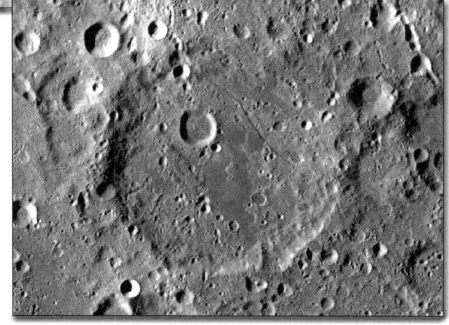

Rima

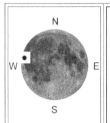

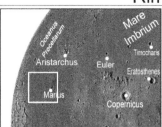

Rima Galilaei	
Lat 12.91N Long 59.2W	
Diameter	185.88Km
Notes:	

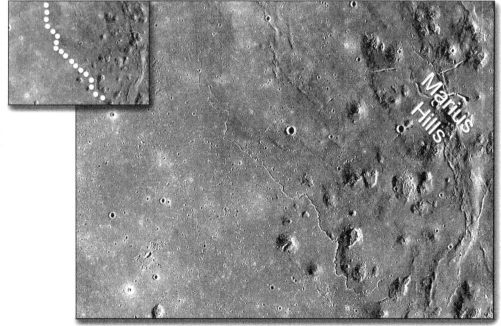

Rima

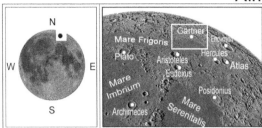

Rima Gärtner	
Lat 58.84N	Long 35.77E
Diameter	42.73Km
Notes:	

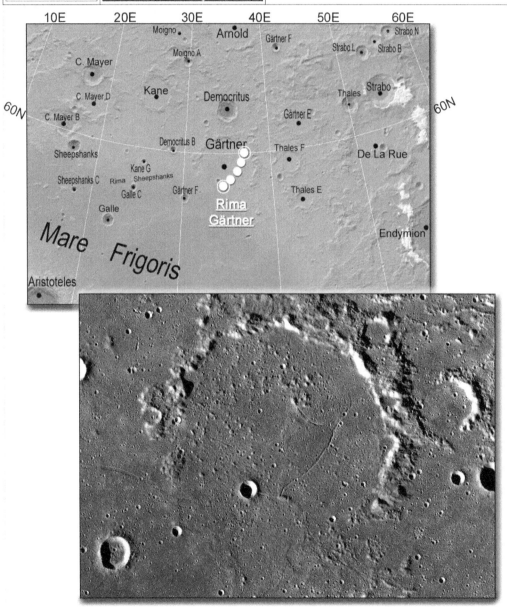

Rima

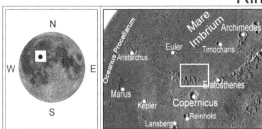

Rima Gay-Lussac	
Lat 13.18N	Long 22.33W
Diameter	40.04Km
Notes:	

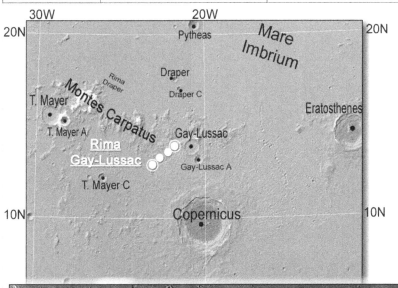

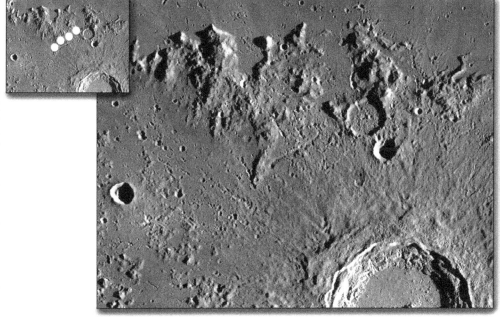

Rima

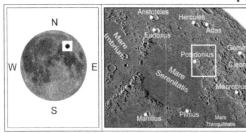

Rima G. Bond	
Lat 32.86N	Long 35.25E
Diameter	166.85Km
Notes:	

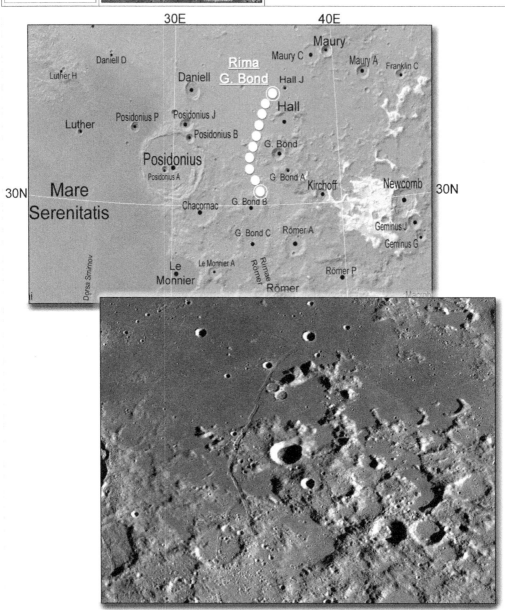

Rima

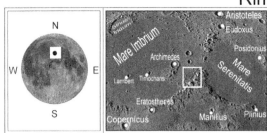

Rima Hadley	
Lat 25.72N	Long 3.15E
Diameter	116.09Km
Notes:	

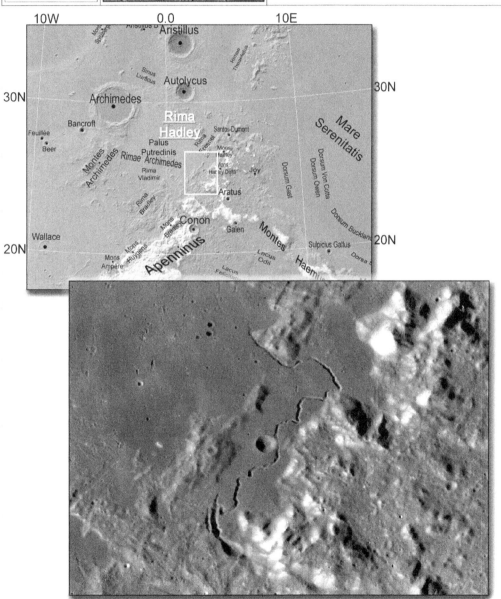

Rima

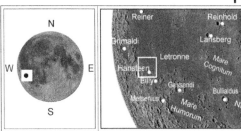

Rima Hansteen	
Lat 12.09S	Long 52.99W
Diameter	30.89Km
Notes:	

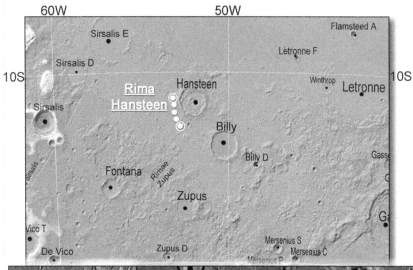

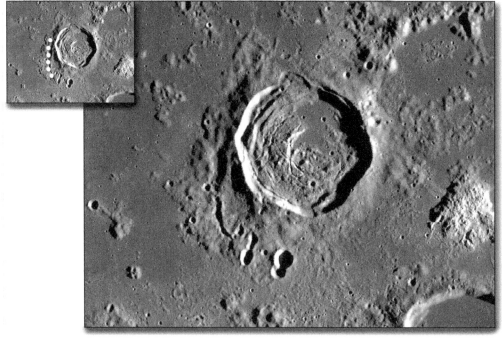

Rima

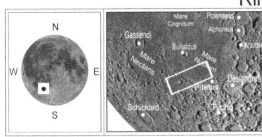

Rima Hesiodus	
Lat 30.54S	Long 21.85W
Diameter	251.46Km
Notes:	

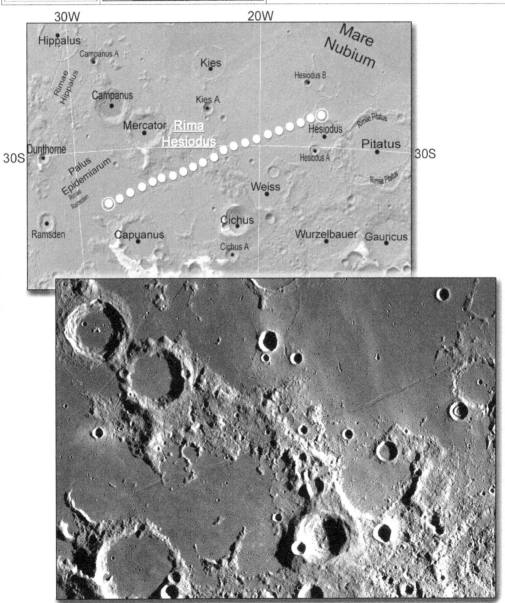

Rima

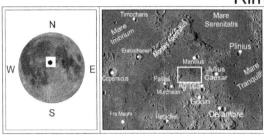

Rima Hyginus	
Lat 7.62N	Long 6.77E
Diameter	203.96Km
Notes:	

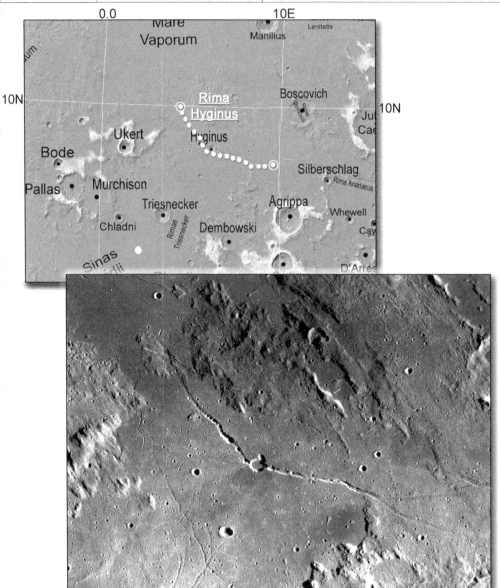

Rima

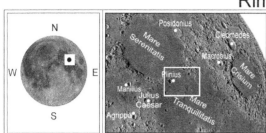

Rima Jansen	
Lat 14.5N	Long 29.51E
Diameter	45.12Km
Notes:	

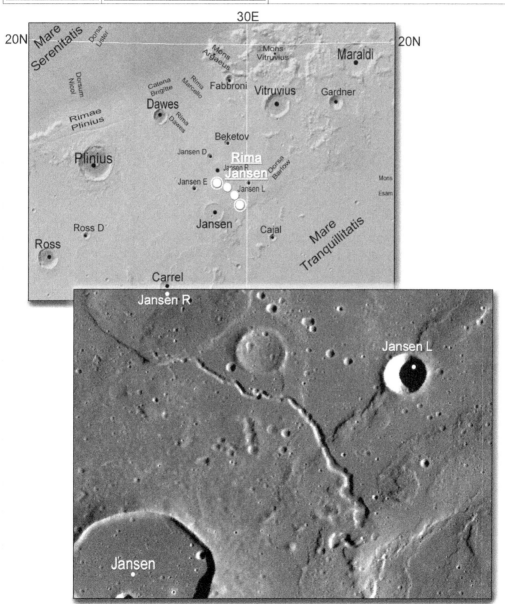

Rima

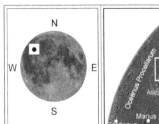

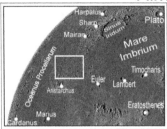

Rima Krieger	
Lat 29.29N	Long 46.26W
Diameter	22.62Km
Notes:	

Rima

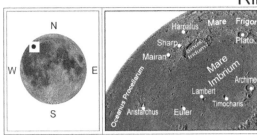

Rima Mairan	
Lat 38.28N Long 46.83W	
Diameter	120.51Km
Notes:	

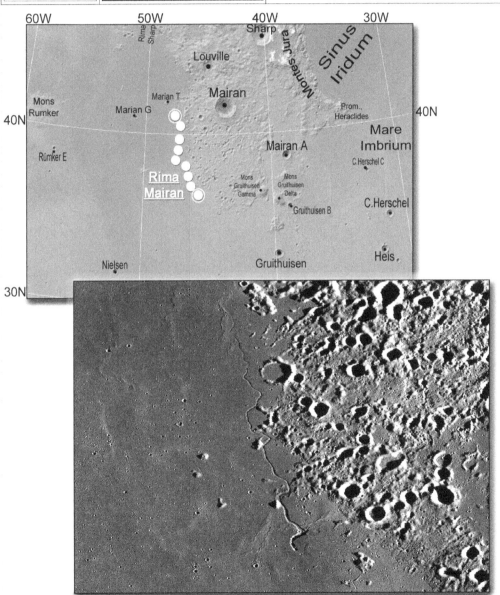

Rima

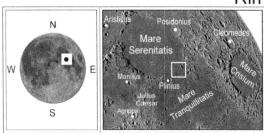

Rima Marcello	
Lat 18.59N	Long 27.74E
Diameter	3.95Km
Notes:	

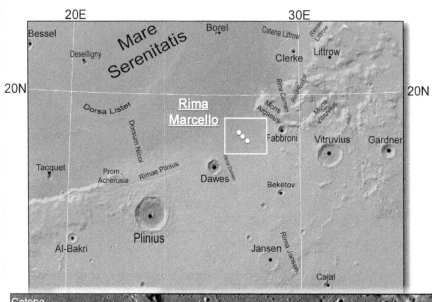

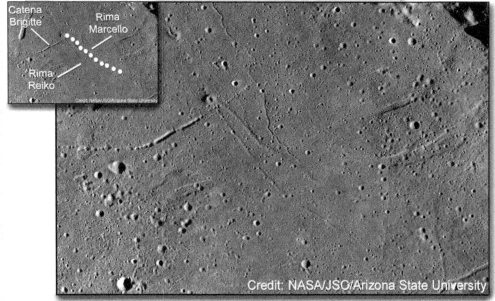

Credit: NASA/JSC/Arizona State University

196

Rima

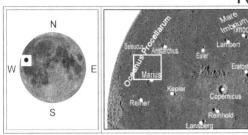

Rima Marius	
Lat 16.37N	Long 49.54W
Diameter	283.54Km
Notes:	

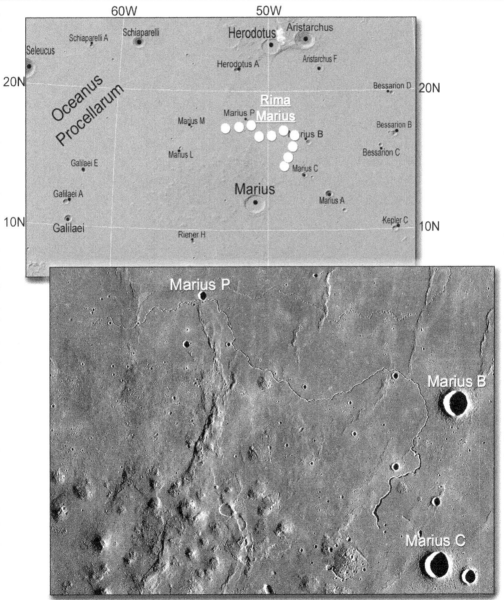

Rima

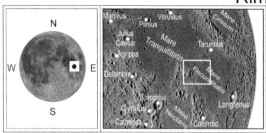

Rima Messier	
Lat 0.76S	Long 44.55E
Diameter	74.22Km
Notes:	

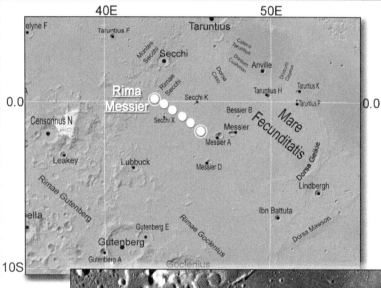

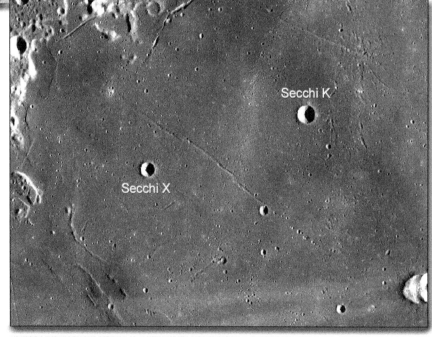

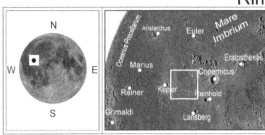

Rima Milichius	
Lat 8.03N	Long 32.87W
Diameter	140.72Km
Notes:	

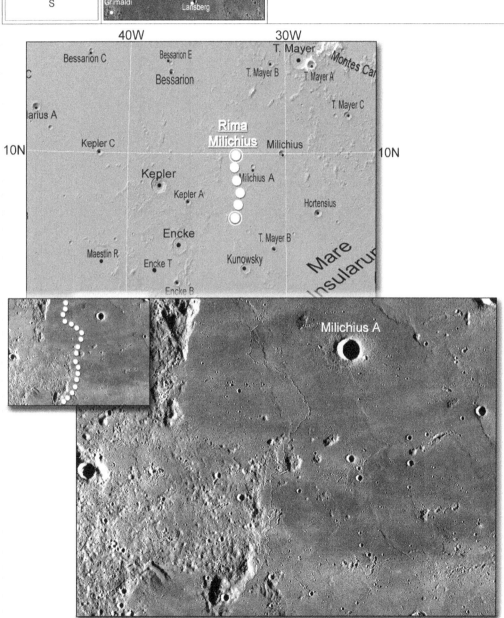

Rima

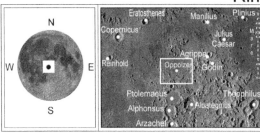

Rima Oppolzer	
Lat 1.53S	Long 1.28E
Diameter	94.2Km
Notes:	

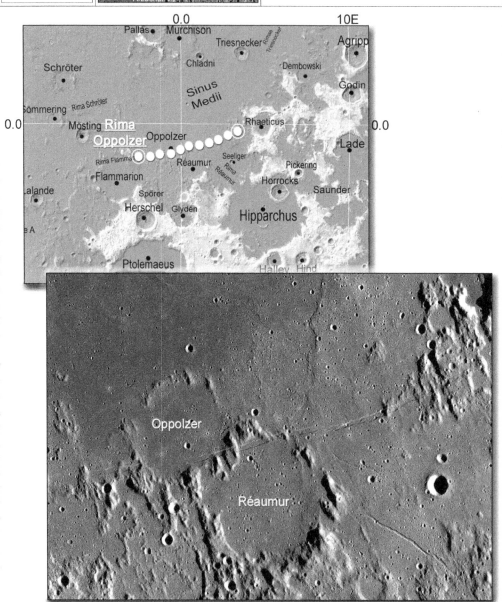

Rima

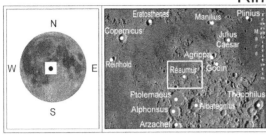

Rima Réaumur	
Lat 2.84S	Long 2.47E
Diameter	30.66Km
Notes:	

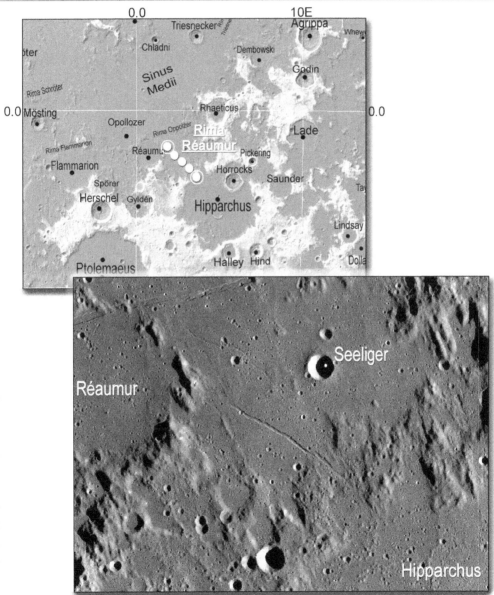

Rima

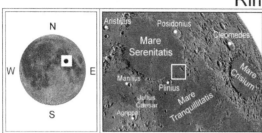

Rima Reiko	
Lat 18.55N	Long 27.71E
Diameter	4.29Km
Notes:	

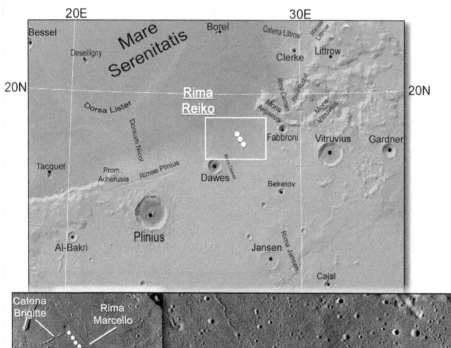

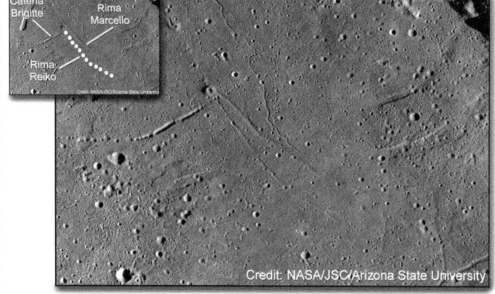

Credit: NASA/JSC/Arizona State University

Rima

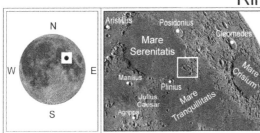

Rima Rudolf	
Lat 19.71N	Long 29.62E
Diameter	8.29Km
Notes:	

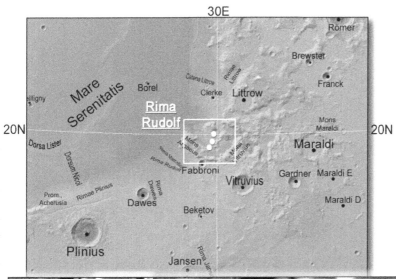

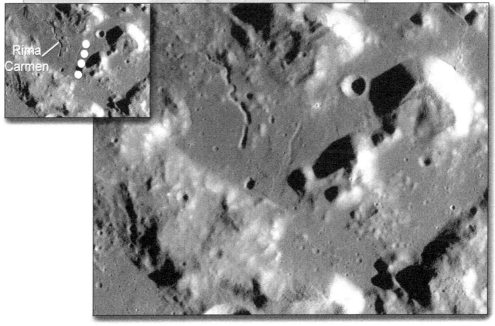

Rima

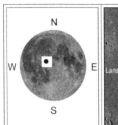

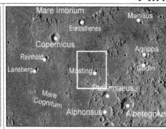

Rima Schröter	
Lat 1.28N	Long 6.25W
Diameter	27.25Km
Notes:	

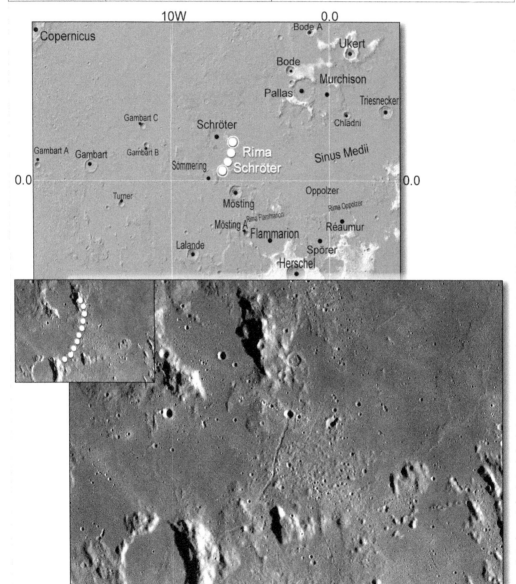

Rima

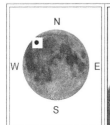

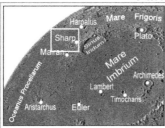

Rima Sharp	
Lat 46.02N	Long 50.36W
Diameter	276.67Km
Notes:	

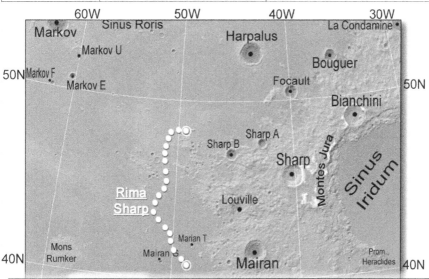

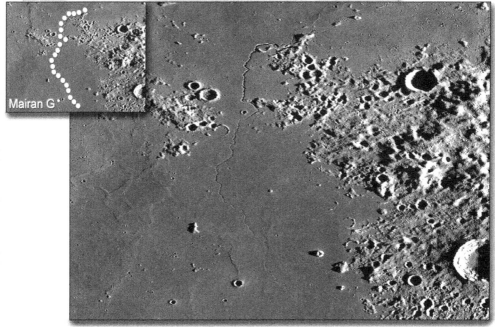

Rima

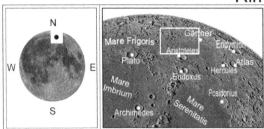

Rima Sheepshanks	
Lat 58.28N	Long 23.69E
Diameter	157.49Km
Notes:	

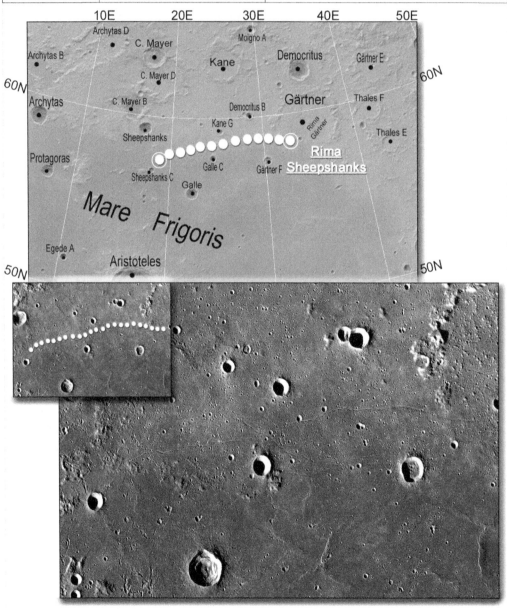

Rima

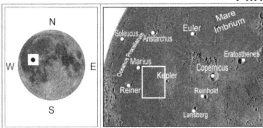

Rima Suess	
Lat 6.62N	Long 47.14W
Diameter	156.39Km
Notes:	

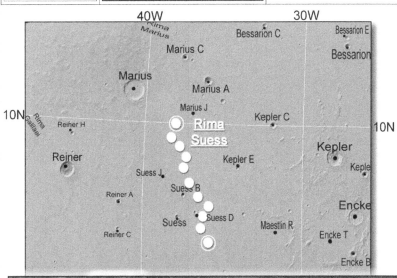

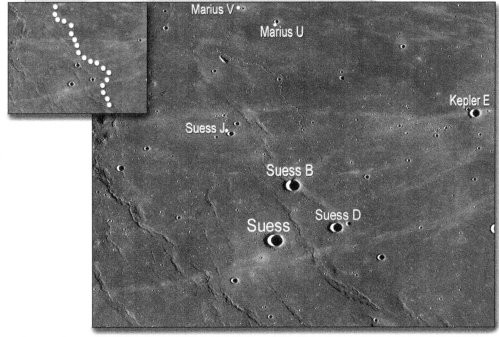

Rima

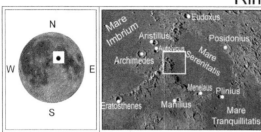

Rima Sung-Mei	
Lat 24.59N	Long 11.28E
Diameter	3.88Km
Notes:	

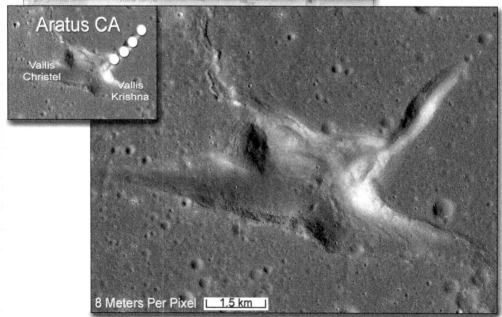

Rima

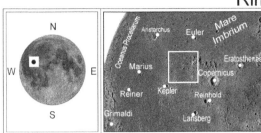

Rima T. Mayer	
Lat 13.26N	Long 31.37W
Diameter	67.81Km
Notes:	

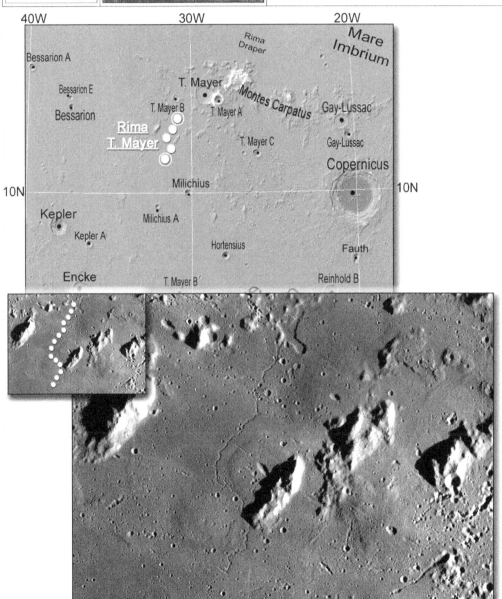

Rima

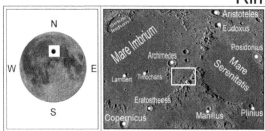

Rima Vladimir	
Lat 25.2N	Long 0.75W
Diameter	10.5Km
Notes:	

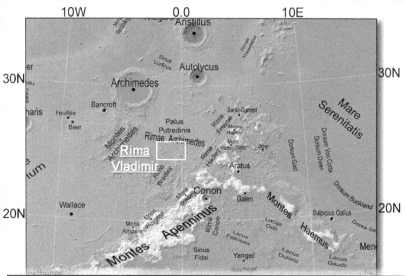

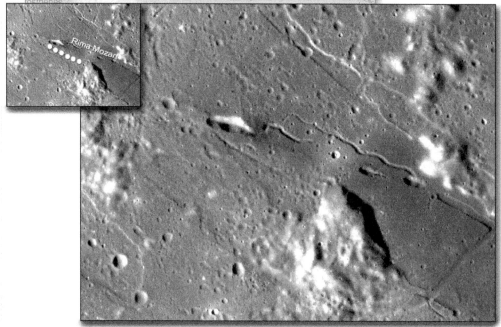

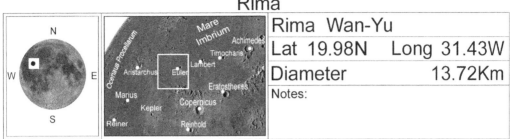

Rima Wan-Yu	
Lat 19.98N	Long 31.43W
Diameter	13.72Km
Notes:	

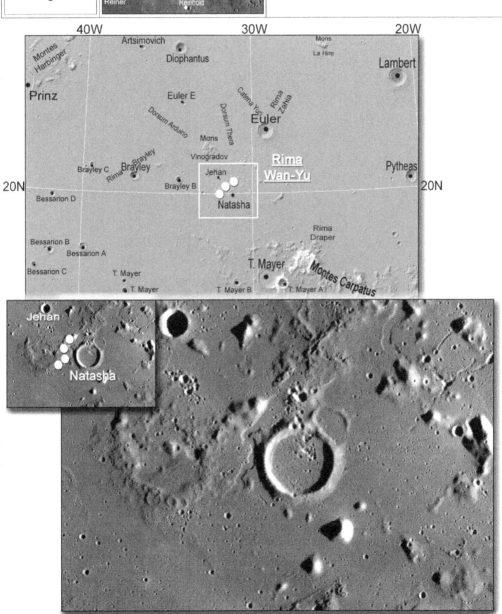

Rima

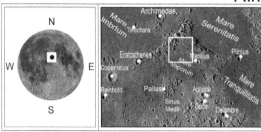

Rima Yangel'	
Lat 16.62N	Long 4.79E
Diameter	30.39Km
Notes:	

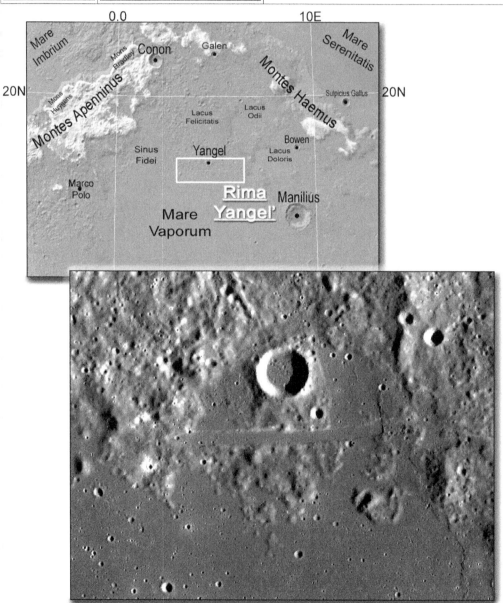

Rima

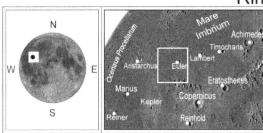

Rima Zahia	
Lat 25.02N	Long 30.46W
Diameter	15.24Km
Notes:	

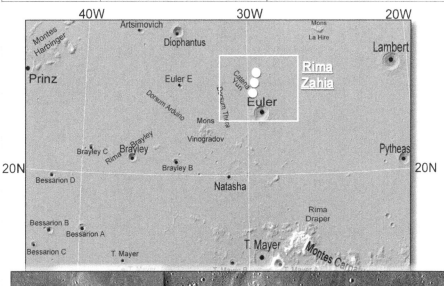

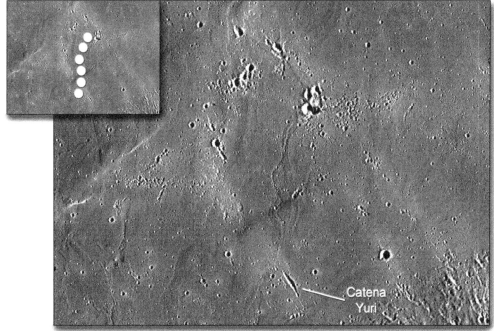

Rimae

Rimae

Rimae is the plural form of the word Rima, which in Latin stands for 'rille' In the lunar context a rille is a volcanically-related feature where lavas, sourced deep from within the Moon's interior, seeped up through fractures onto the surface. In effect, the lavas channelled up and through along huge tubes, which after emptying a long time after (in some cases upto millions of years later) led to a collapsed cavity on the lunar surface. Some of these cavities (rilles) can form into very long linear-like features e.g. Rima Ariadaeus, while others can be very sinuous-like e.g. Rima Marius (see the Rima pages for more).

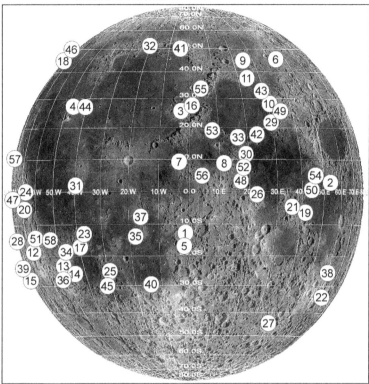

No.	Feature	No.	Feature	No.	Feature
1	Alphonsus	21	Gutenberg	41	Plato
2	Apollonius	22	Hase	42	Plinius
3	Archimedes	23	Herigonius	43	Posidonius
4	Aristarchus	24	Hevelius	44	Prinz
5	Arzachel	25	Hippalus	45	Ramsden
6	Atlas	26	Hypatia	46	Repsold
7	Bode	27	Janssen	47	Riccioli
8	Boscovich	28	Kopff	48	Ritter
9	Bürg	29	Littrow	49	Römer
10	Chacornac	30	Maclear	50	Secchi
11	Daniell	31	Maestlin	51	Sirsalis
12	Darwin	32	Maupertuis	52	Sosigenes
13	da Gasparis	33	Menelaus	53	Sulpicius Gallus
14	Doppelmayer	34	Mersenius	54	Taruntius
15	Focas	35	Opelt	55	Theaetetus
16	Fresnel	36	Palmieri	56	Triesnecker
17	Gassendi	37	Parry	57	Vasco da Gamma
18	Gerard	38	Petavius	58	Zupus
19	Goclenius	39	Pettit		
20	Grimaldi	40	Pitatus		

Note: The above list of Rimae represents those given in the International Astronomical Union list, but there are plenty more on the Nearside's face that may in the future be given official designations.

Rimae

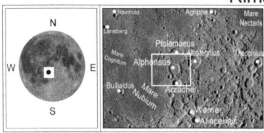

Rimae Alphonsus	
Lat 13.4S	Long 1.94W
Diameter	87.0Km
Notes:	

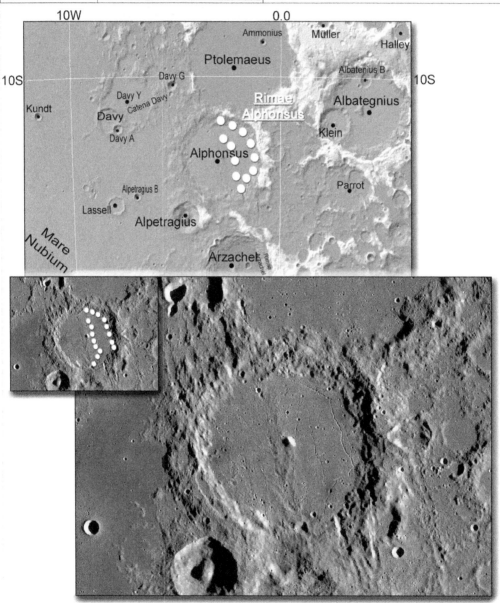

Rimae

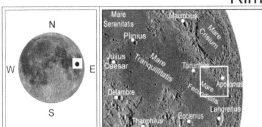

Rimae Apollonius	
Lat 4.39N	Long 54.33E
Diameter	89.64Km
Notes:	

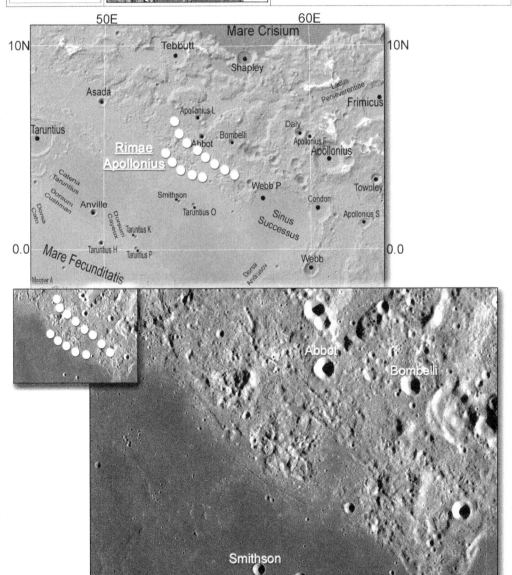

Rimae

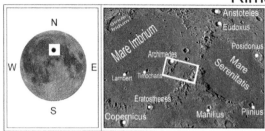

Rimae Archimedes	
Lat 26.34N	Long 4.53W
Diameter	215.0Km
Notes:	

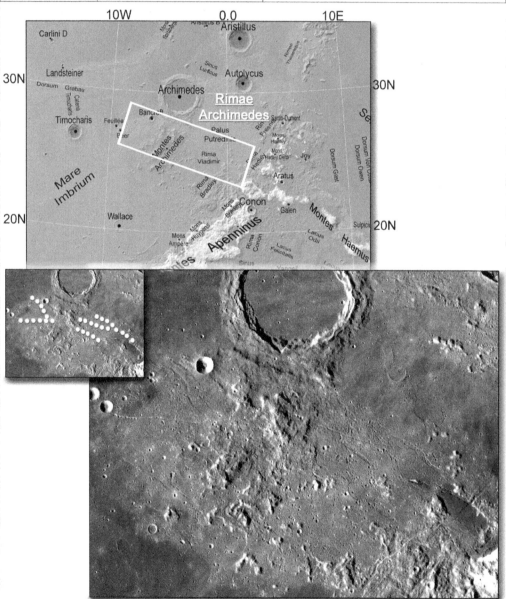

218

Rimae

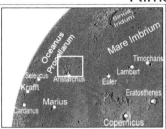

Rimae Aristarchus	
Lat 27.52N	Long 47.25W
Diameter	175.0Km
Notes:	

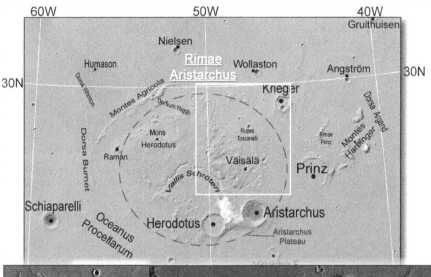

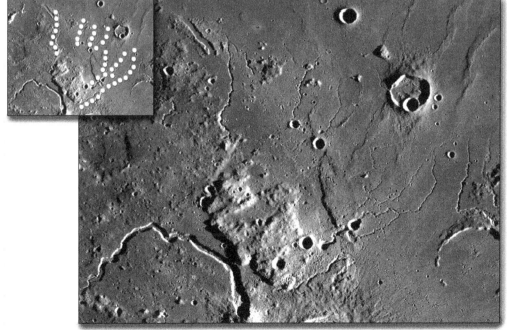

Rimae

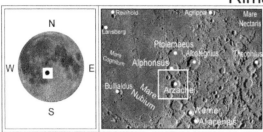

Rimae Arzachel	
Lat 18.31S	Long 1.38W
Diameter	57.0Km
Notes:	

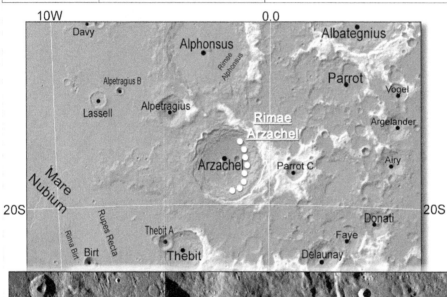

Rimae

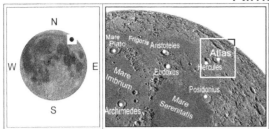

Rimae Atlas	
Lat 46.82N	Long 44.42E
Diameter	46.8Km
Notes:	

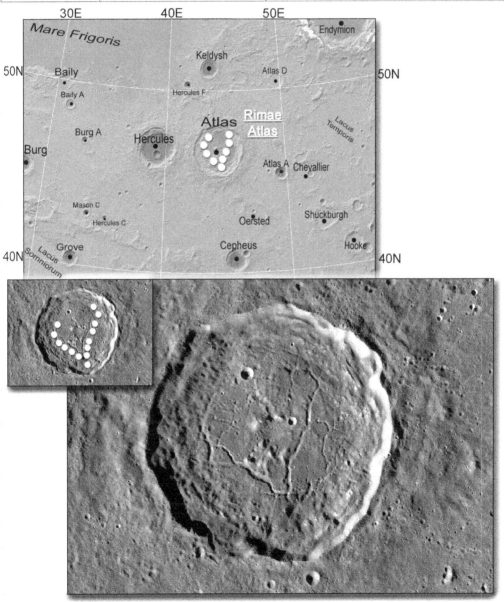

Rimae

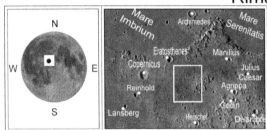

Rimae Bode	
Lat 9.54N	Long 3.22W
Diameter	233.0Km
Notes:	

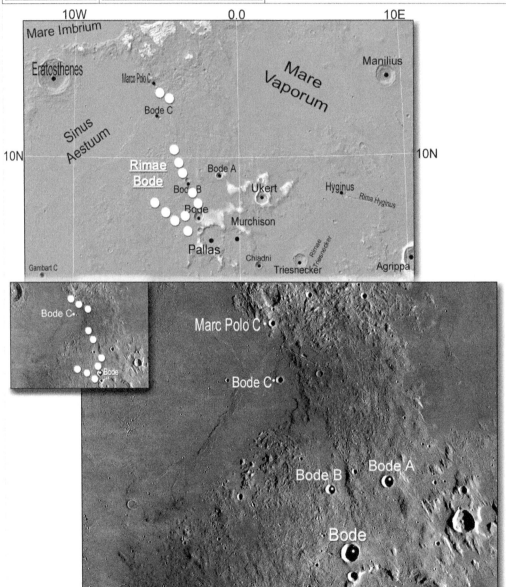

Rimae

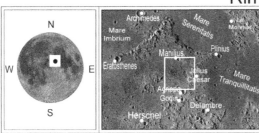

Rimae Boscovich	
Lat 9.87N	Long 11.27E
Diameter	32.0Km
Notes:	

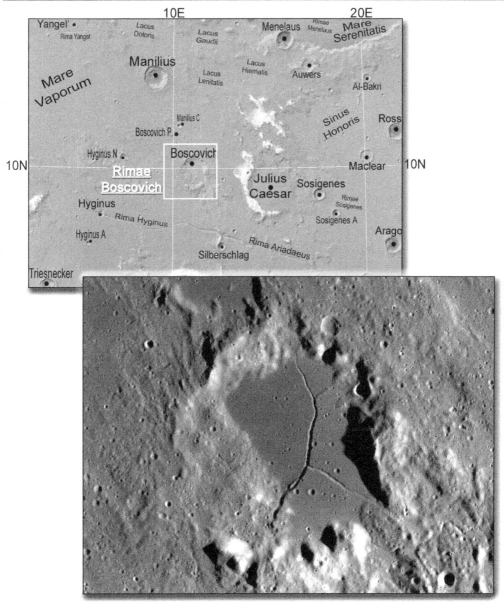

Rimae

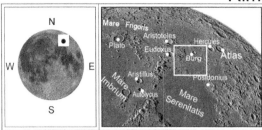

Rimae Bürg	
Lat 44.7N	Long 25.27E
Diameter	98.0Km
Notes:	

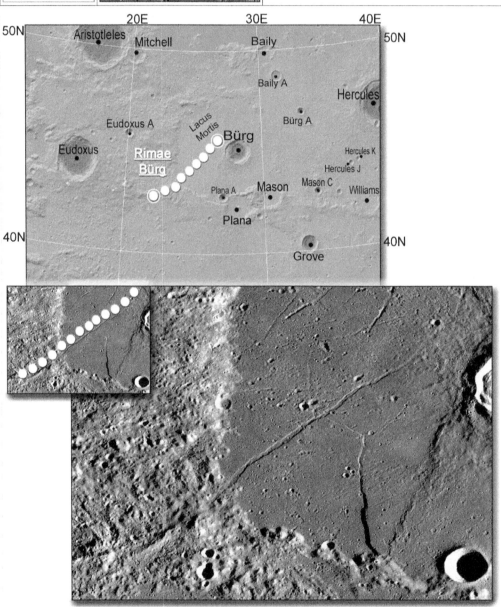

224

Rimae

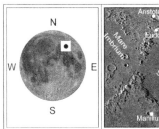

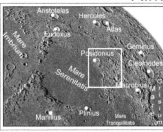

Rimae Chacornac	
Lat 29.01N	Long 31.24E
Diameter	100.0Km
Notes:	

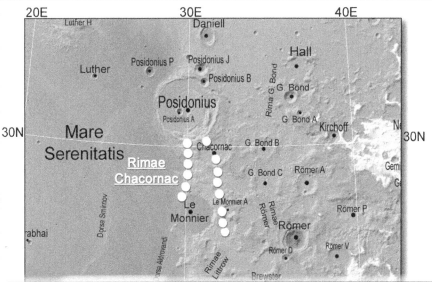

Rimae

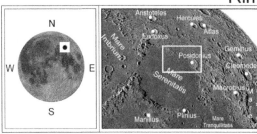

Rimae Daniell	
Lat 37.53N	Long 24.33E
Diameter	140.25Km
Notes:	

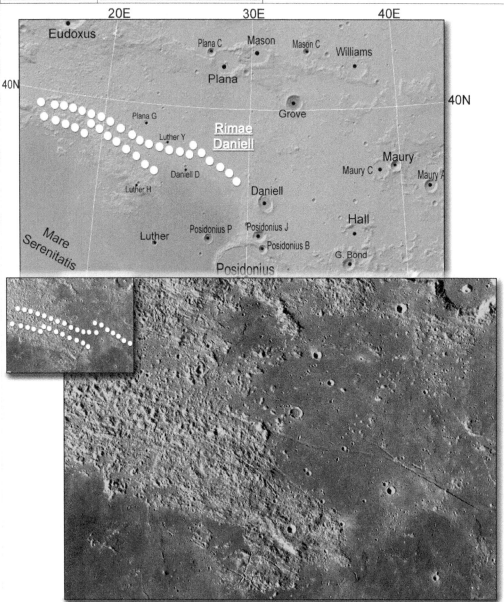

Rimae

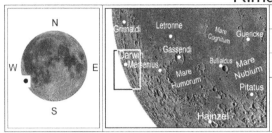

Rimae Darwin	
Lat 19.84S	Long 66.66W
Diameter	170.0Km
Notes:	

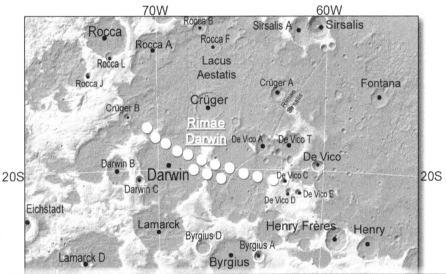

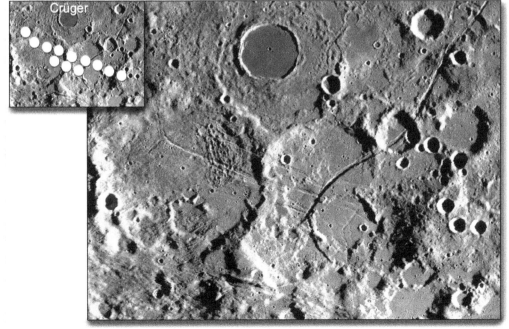

Rimae

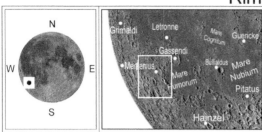

Rimae de Gasparis	
Lat 24.99S	Long 50.3W
Diameter	46.65Km
Notes:	

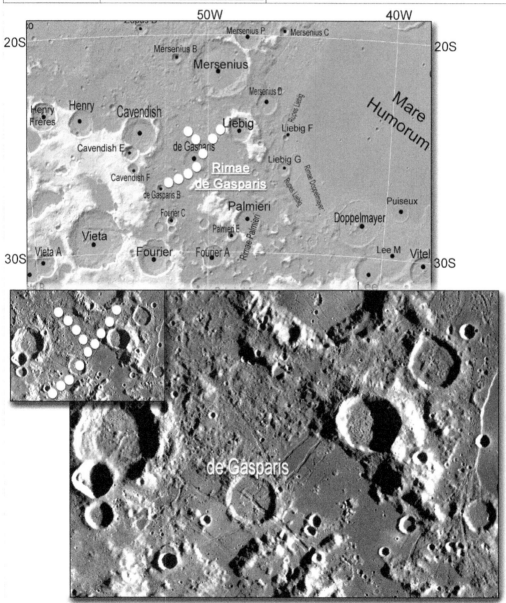

Rimae

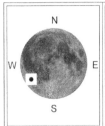

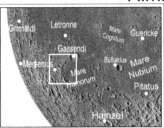

Rimae Doppelmayer	
Lat 26.23S	Long 44.53W
Diameter	7.8Km
Notes:	

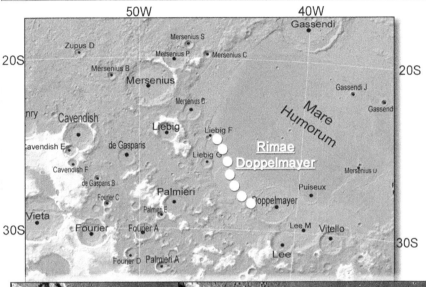

Rimae

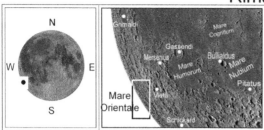

Rimae Focas	
Lat 27.68S	**Long 97.54W**
Diameter	**61.0Km**
Notes:	

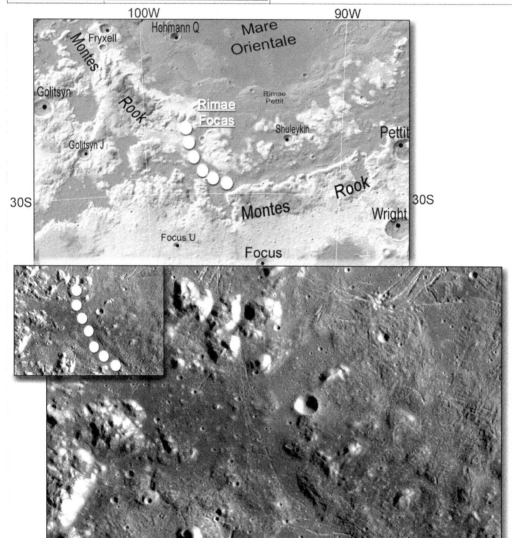

Rimae

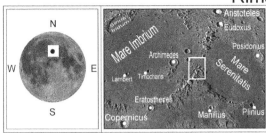

Rimae Fresnel	
Lat 28.11N	Long 3.73E
Diameter	75.0Km
Notes:	

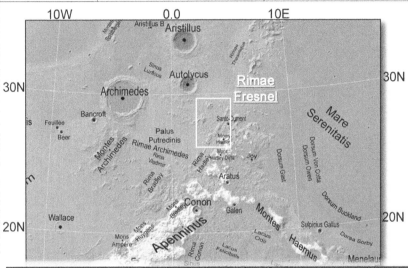

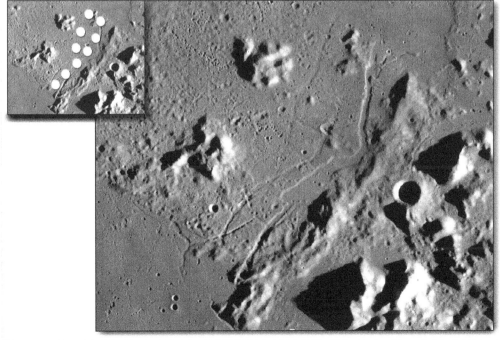

Rimae

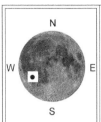

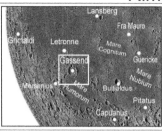

Rimae Gassendi	
Lat 17.47S	Long 39.87W
Diameter	70.0Km
Notes:	

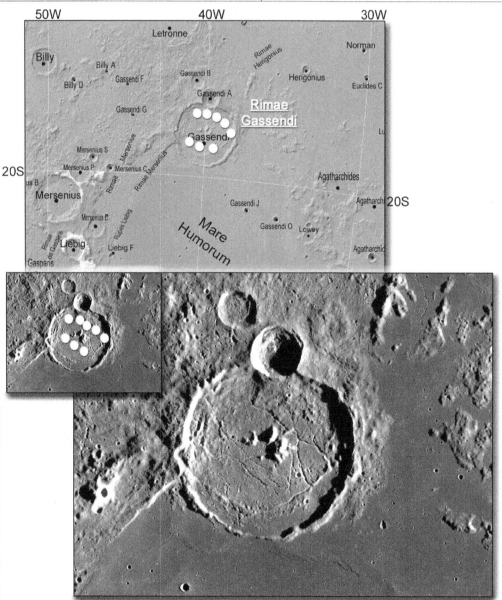

Rimae

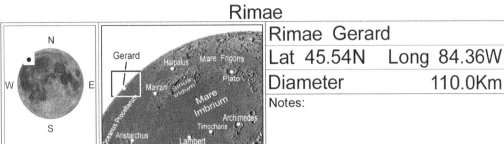

Rimae Gerard	
Lat 45.54N	Long 84.36W
Diameter	110.0Km
Notes:	

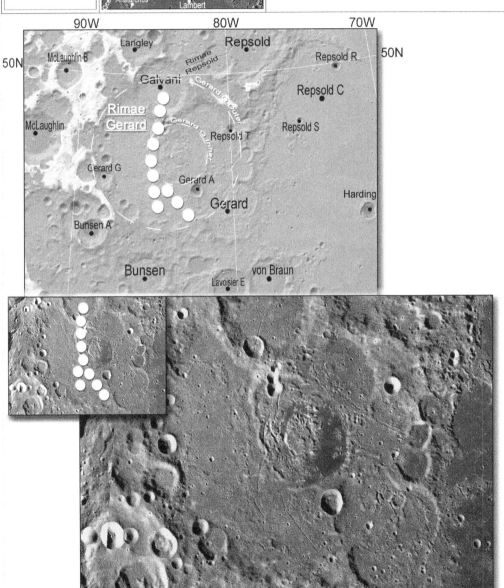

Rimae

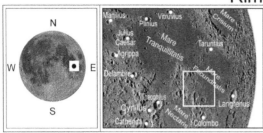

Rimae Goclenius	
Lat 7.84S	Long 42.88E
Diameter	190.0Km
Notes:	

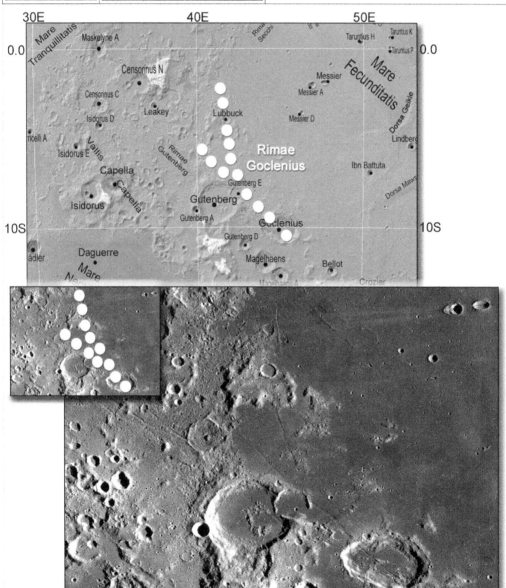

Rimae

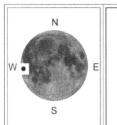

Rimae Grimaldi	
Lat 6.18S	Long 63.9W
Diameter	162.0Km
Notes:	

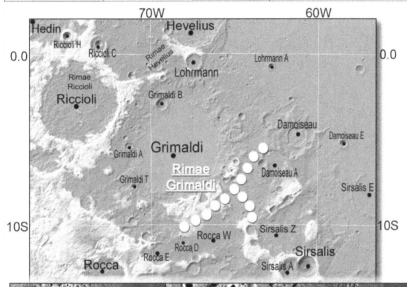

Rimae

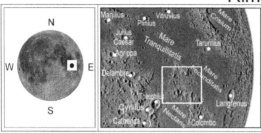

Rimae Gutenberg

Lat 4.42S	Long 36.42E
Diameter	223.0Km

Notes:

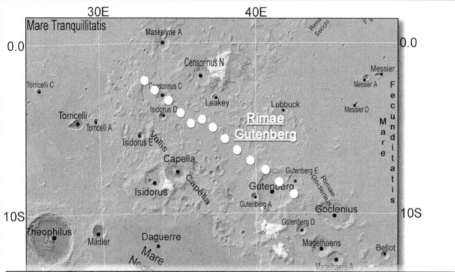

Mare Tranquillitatis

30E   40E

Rima Snooti

0.0   Maskelyne A   0.0

Censorinus N

Torricelli C   Censorinus C   Leakey   Lubbuck   Messier   Messier A

Isidorus D   Messier D

Torricelli   **Rimae**

Torricelli A   **Gutenberg**

Isidorus E   Vallis   Mare

Capella   Gutenberg E   Fecunditatis

Isidorus   Capella   Gutenberg   Rima Goclenius

Gutenberg A

Goclenius

10S   Theophilus   Gutenberg D   10S

Mädler   Daguerre   Magelhaens   Bellot

Mare Nect   Magelhaens A

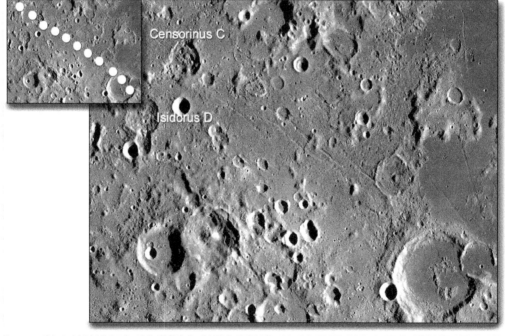

Censorinus C

Isidorus D

Rimae

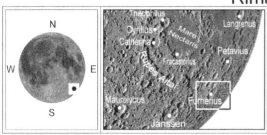

Rimae Hase	
Lat 34.71S	Long 67.78E
Diameter	257.24Km
Notes:	

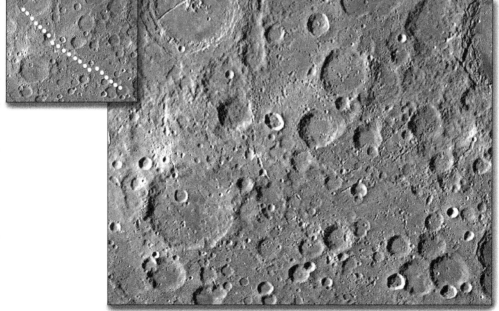

60E   70E

Petavius
Palitzsch A   Palitzsch B
Phillips
Possible Extension
Vallis Palitzsch
Hase
Legendre
Legendre K
30S   30S
Vallis Snellius
Adams B
Adams
Legendre D
Furnerius C
Adams D
Rimae Hase
Rima Furnerius
Furnerius J
Furnerius B
Adams P.
Furnerius V
Furnerius
Furnerius G   Furnerius H
Marinus C.
Fraunhofer
Furnerius K
Marinus B
40S   Marinus 40S

Rimae

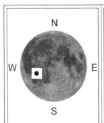

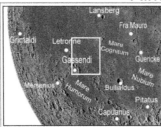

Rimae Herigonius	
Lat 13.92S	Long 36.75W
Diameter	180.0Km
Notes:	

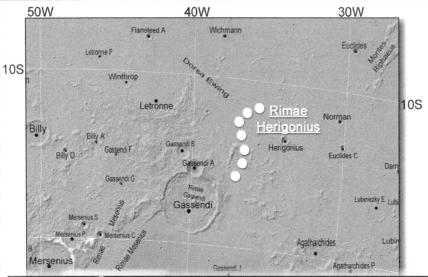

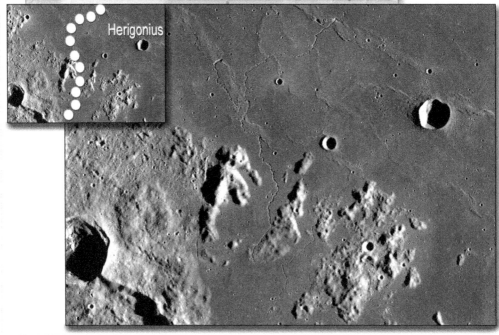

Rimae

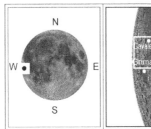

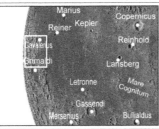

Rimae Hevelius	
Lat 0.81N	Long 66.38W
Diameter	180.0Km
Notes:	

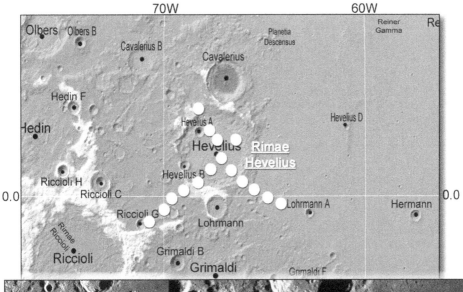

Rimae

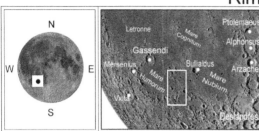

Rimae Hippalus	
Lat 25.6S	Long 29.36W
Diameter	266.0Km
Notes:	

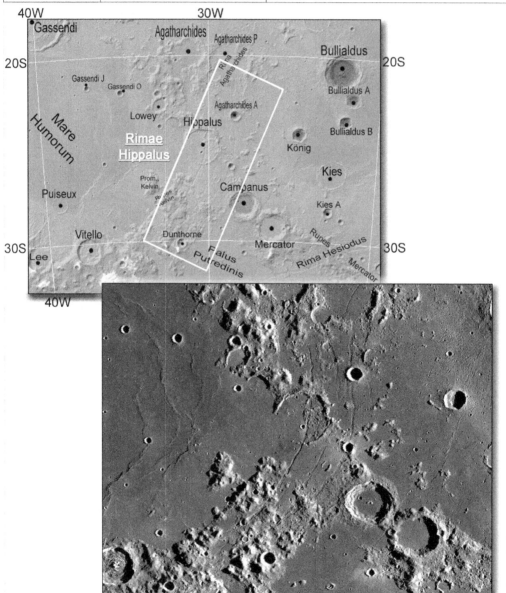

Rimae

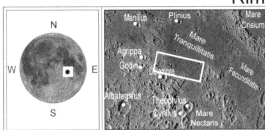

Rimae Hypatia	
Lat 0.34S	Long 22.78E
Diameter	200.0Km
Notes:	

20E 30E

Rima Ariadaeus

Mare Tranquillitatis

Arago
Lamont

Whewell Cayley

Manners

Rima Ritter

D'Arrest

Ritter

Dionysius

Maskelyne

Rimae
Hypatia

0.0 Sabine 0.0

Theon Jnr.,

Delambre

Torricelli C

Theon Snr.,

Hypatia

Taylor A

Sinus
Asperitatis

Torricelli

Taylor

Rimae

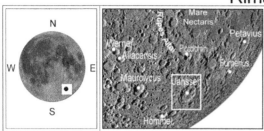

Rimae Janssen	
Lat 45.8S	Long 39.26E
Diameter	120.0Km
Notes:	

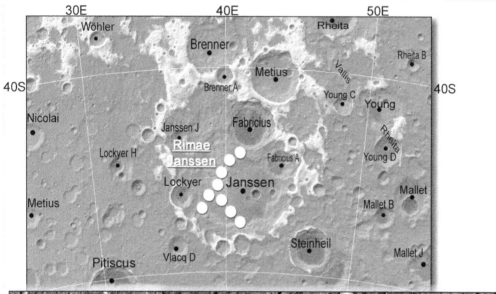

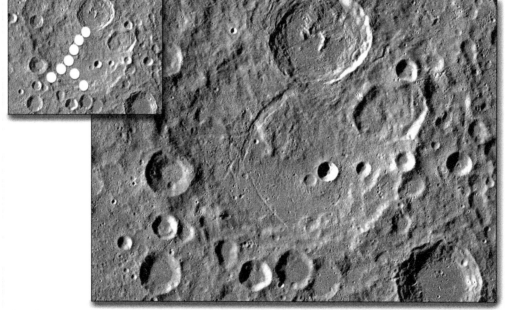

Rimae

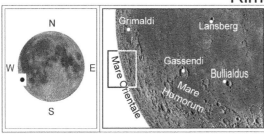

Rimae Kopff	
Lat 14.68S	Long 88.1W
Diameter	250.0Km
Notes:	

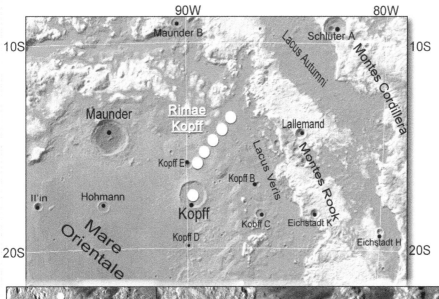

Rimae

Rimae Littrow	
Lat 22.47N	Long 30.47E
Diameter	165.0Km
Notes:	

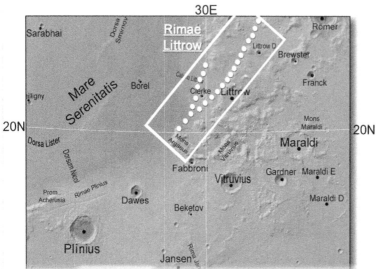

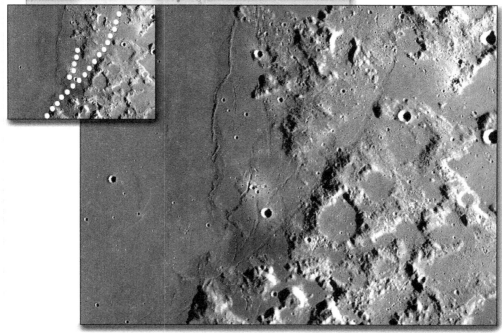

Rimae

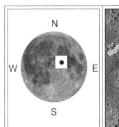

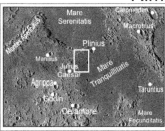

Rimae Maclear	
Lat 12.23N	Long 19.9E
Diameter	94.0Km
Notes:	

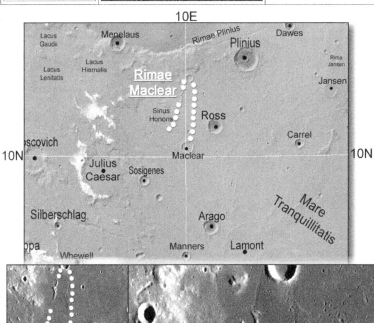

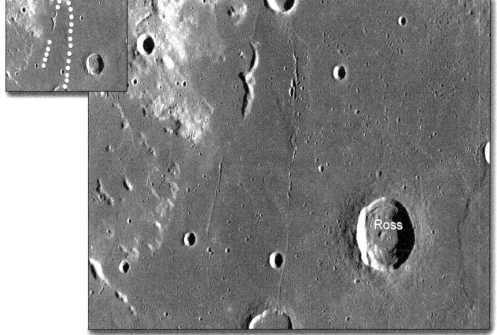

Rimae

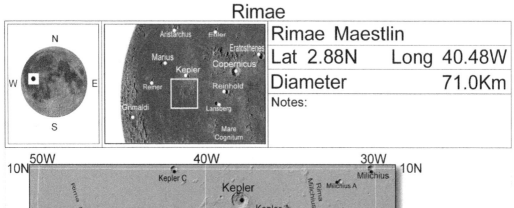

Rimae Maestlin	
Lat 2.88N	Long 40.48W
Diameter	71.0Km
Notes:	

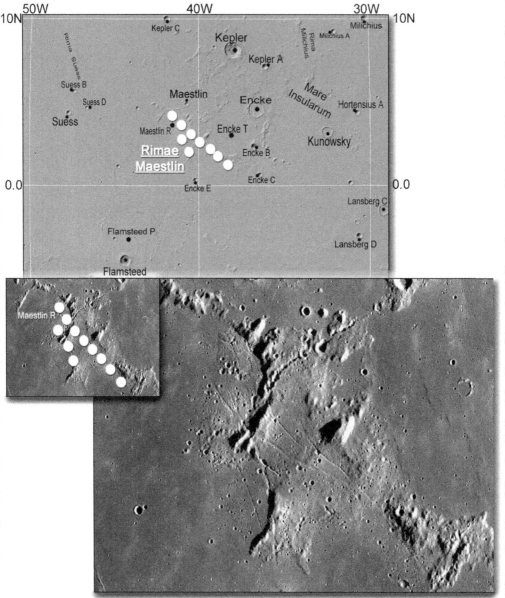

Rimae

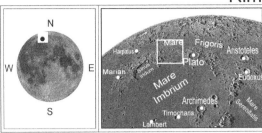

Rimae Maupertuis	
Lat 51.24N	Long 22.82W
Diameter	50.0Km
Notes:	

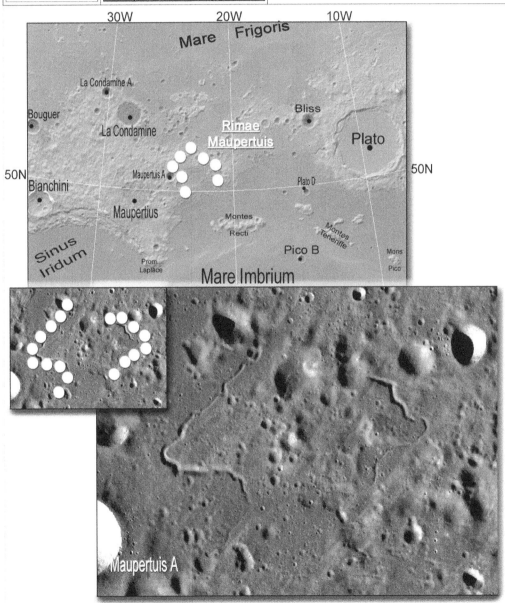

Rimae

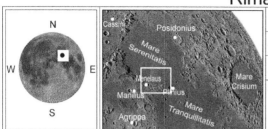

Rimae Menelaus	
Lat 17.1N	Long 17.77E
Diameter	87.0Km
Notes:	

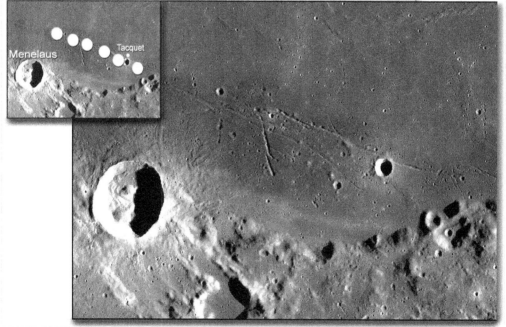

Rimae

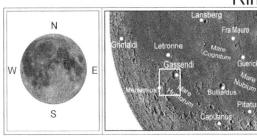

Rimae Mersenius	
Lat 20.69S	Long 46.53W
Diameter	240.0Km
Notes:	

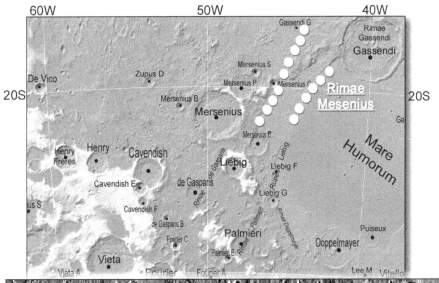

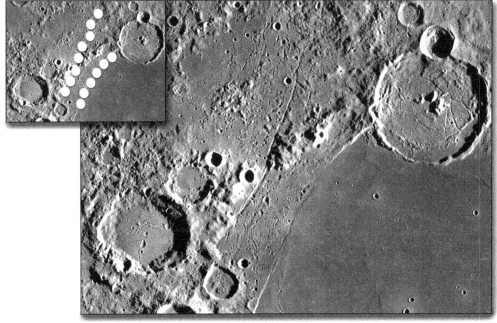

Rimae

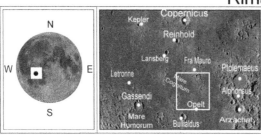	**Rimae Opelt**	
	Lat 13.64S	Long 18.14W
	Diameter	55.0Km
	Notes:	

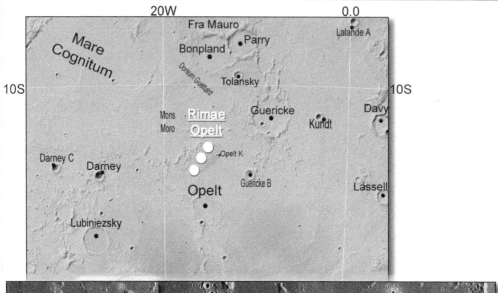

Rimae

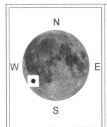

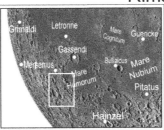

Rimae Palmieri	
Lat 27.83S	Long 47.17W
Diameter	27.13Km
Notes:	

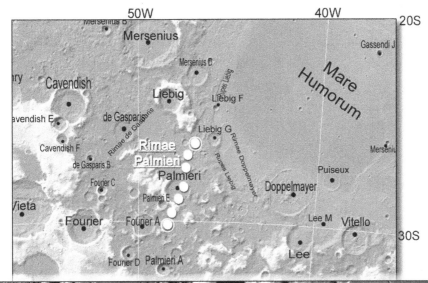

Rimae

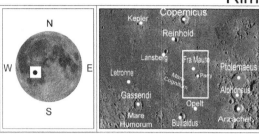

Rimae Parry	
Lat 8.07S	Long 16.52W
Diameter	210.0Km
Notes:	

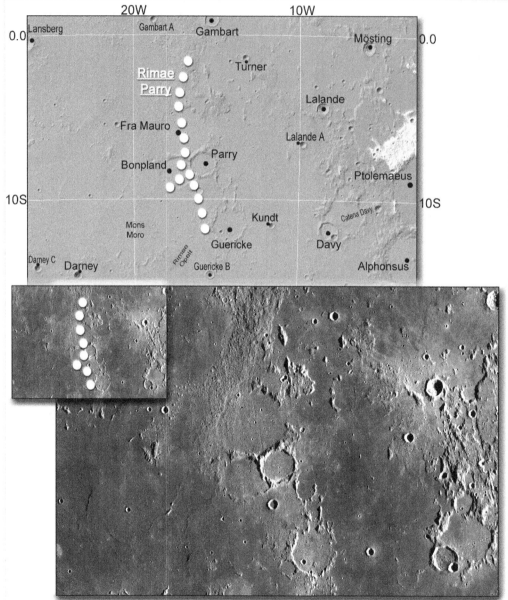

Rimae

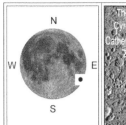

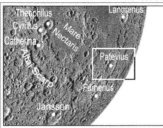

Rimae Petavius	
Lat 25.23S	Long 60.48E
Diameter	110.0Km
Notes:	

Rimae

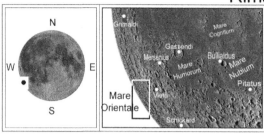

Rimae Pettit	
Lat 25.22S	Long 93.63W
Diameter	320.0Km
Notes:	

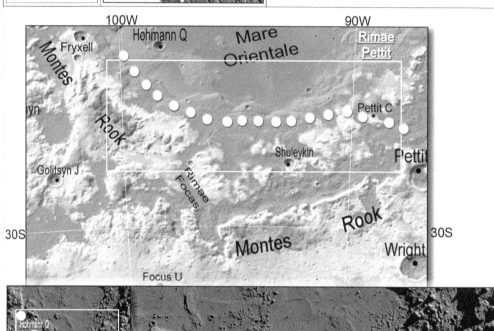

Rimae

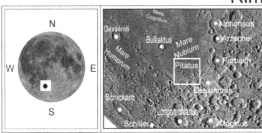

Rimae Pitatus	
Lat 29.84S	Long 13.62W
Diameter	90.0Km
Notes:	

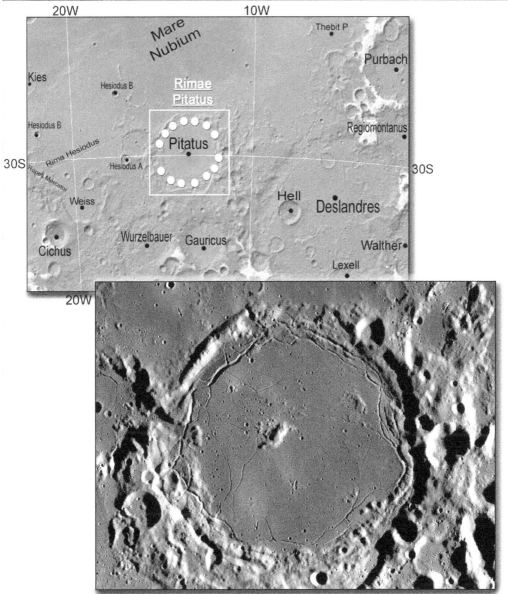

Rimae

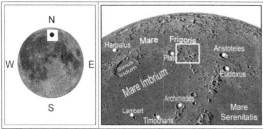

Rimae Plato	
Lat 50.88N	Long 3.02W
Diameter	180.0Km
Notes:	

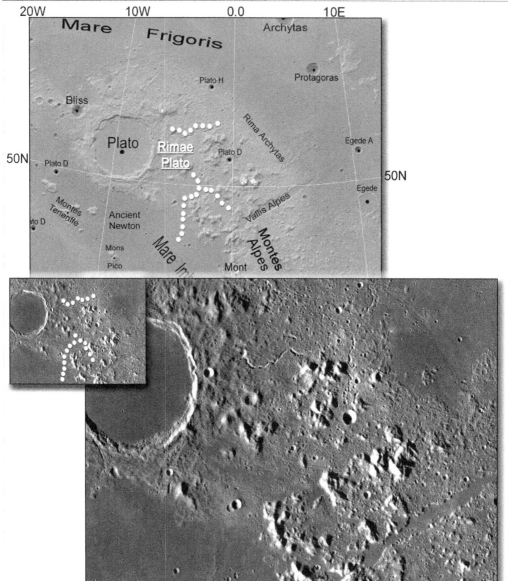

Rimae

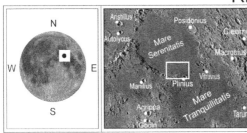

Rimae Plinius	
Lat 17.05N	Long 23.14E
Diameter	100.0Km
Notes:	

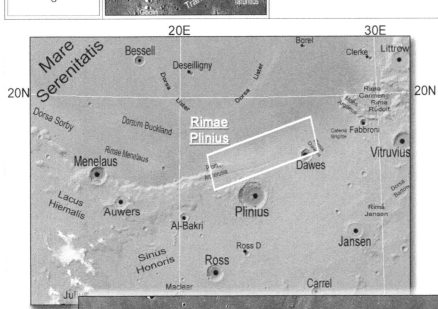

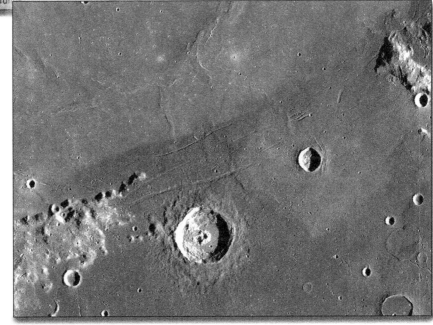

Rimae

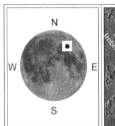

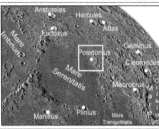

Rimae Posidonius	
Lat 32.03N	Long 29.61E
Diameter	78.0Km
Notes:	

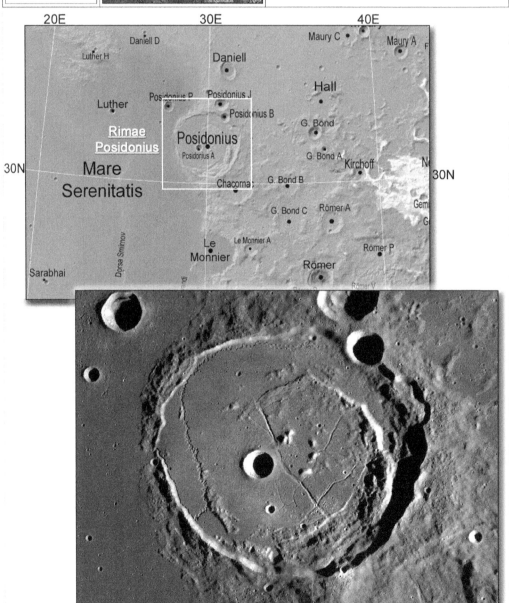

Rimae

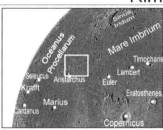

Rimae Prinz	
Lat 27.05N	Long 43.51W
Diameter	10.95Km
Notes:	

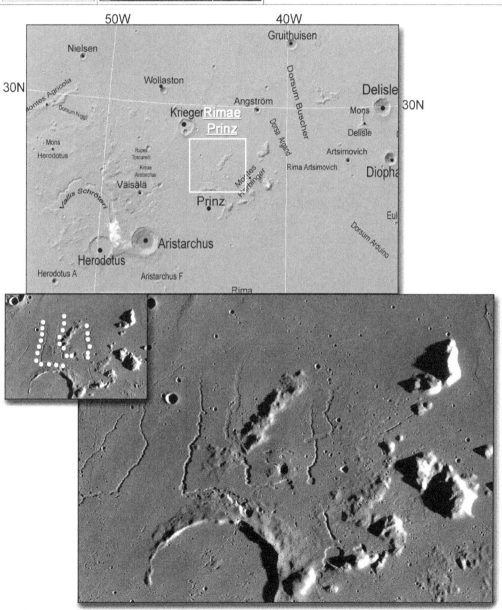

Rimae

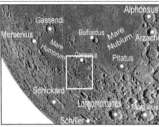

Rimae Ramsden	
Lat 32.93S	Long 31.32W
Diameter	100.0Km
Notes:	

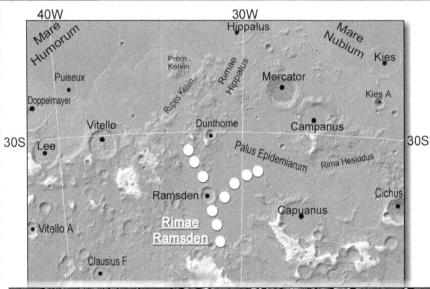

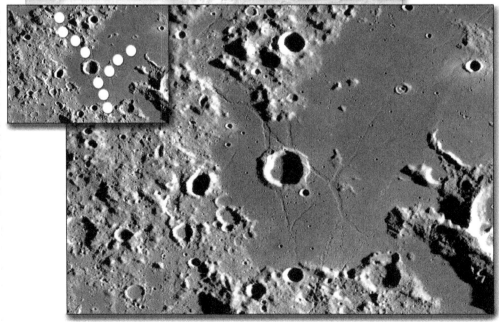

Rimae

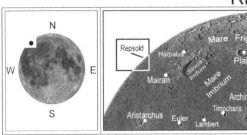

Rimae Repsold	
Lat 50.74N	Long 80.46W
Diameter	152.05Km
Notes:	

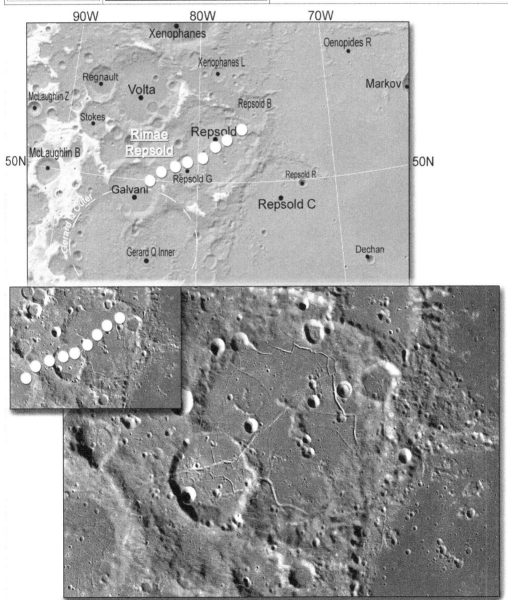

Rimae

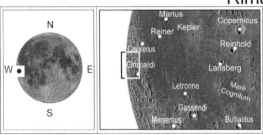

Rimae Riccioli	
Lat 1.52S	Long 73.07W
Diameter	250.0Km
Notes:	

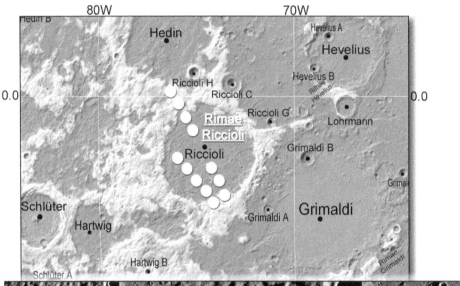

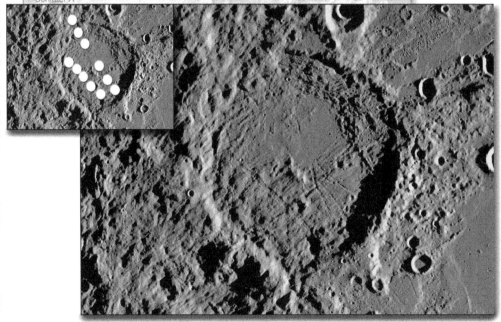

Rimae

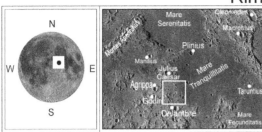

Rimae Ritter	
Lat 3.5N	Long 17.97E
Diameter	75.0Km
Notes:	

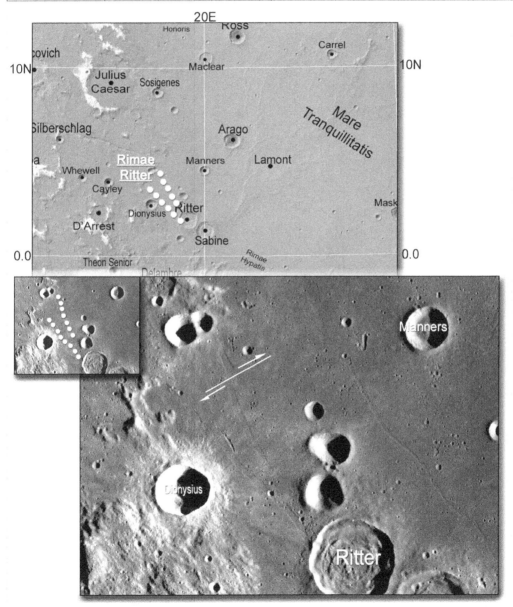

263

Rimae

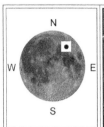

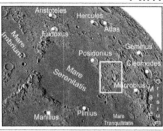

Rimae Römer	
Lat 26.98N	Long 34.86E
Diameter	112.0Km
Notes:	

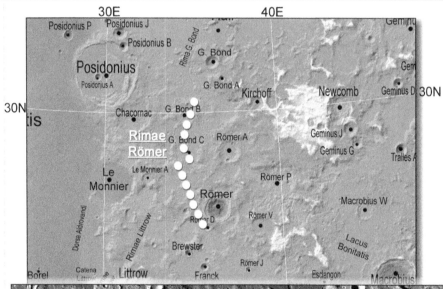

Rimae

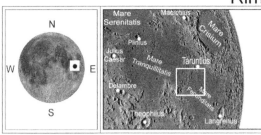

Rimae Secchi	
Lat 0.99N	Long 44.08E
Diameter	35.0Km
Notes:	

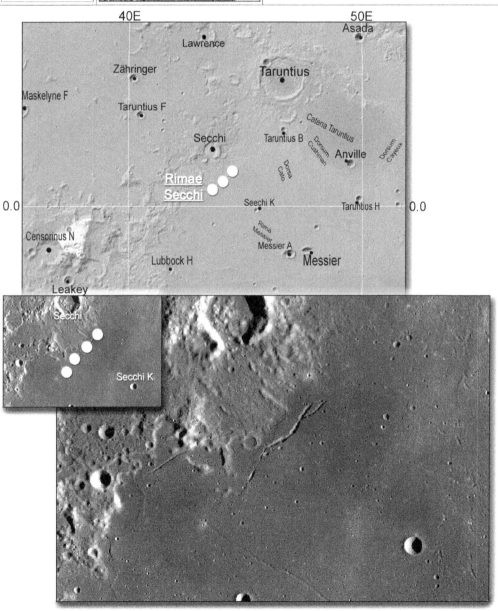

Rimae

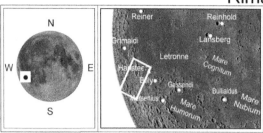

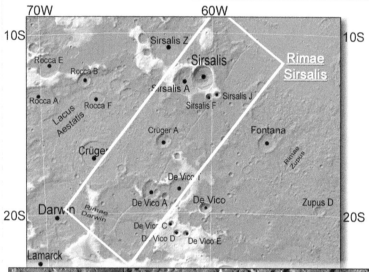

Rimae

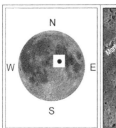

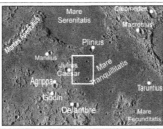

Rimae Sosigenes	
Lat 8.08N	Long 18.72E
Diameter	130.0Km
Notes:	

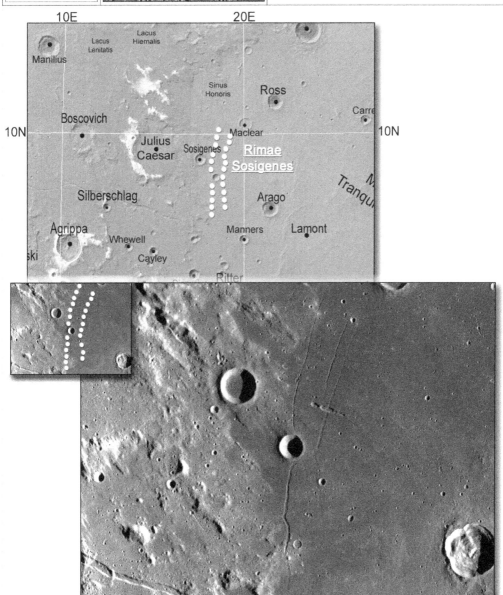

Rimae

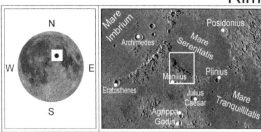

Rimae Sulpicius Gallus	
Lat 20.65N	Long 9.99E
Diameter	80.0Km
Notes:	

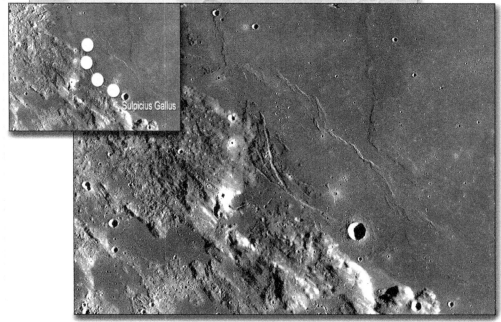

Rimae

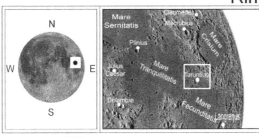

Rimae Taruntius	
Lat 5.83N	Long 46.83E
Diameter	35.0Km
Notes:	

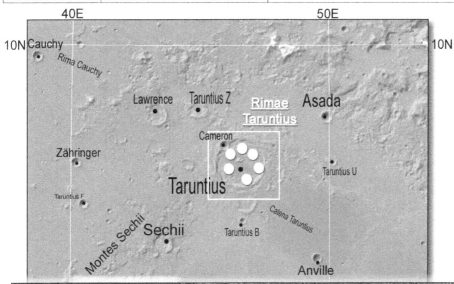

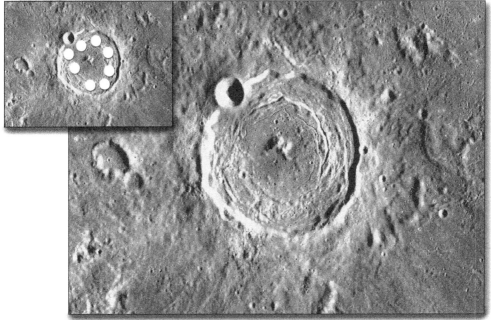

Rimae

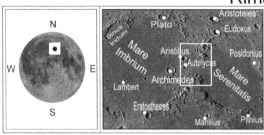

Rimae Theaetetus	
Lat 33.04N	Long 5.87E
Diameter	53.0Km
Notes:	

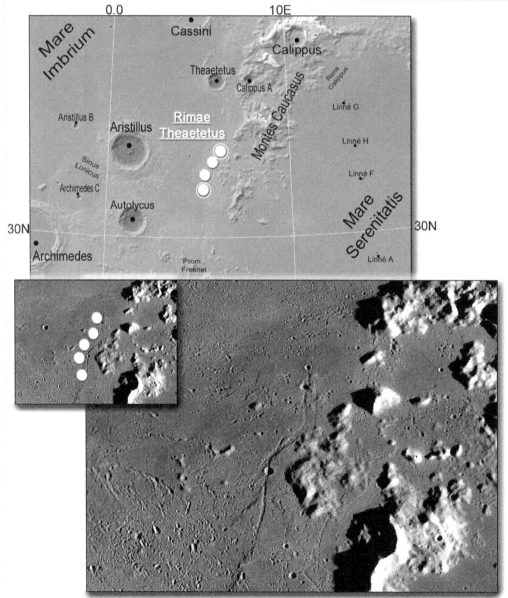

Rimae

	Rimae Triesnecker	
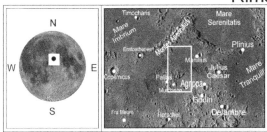	Lat 5.1N	Long 4.83E
	Diameter	200.0Km
	Notes:	

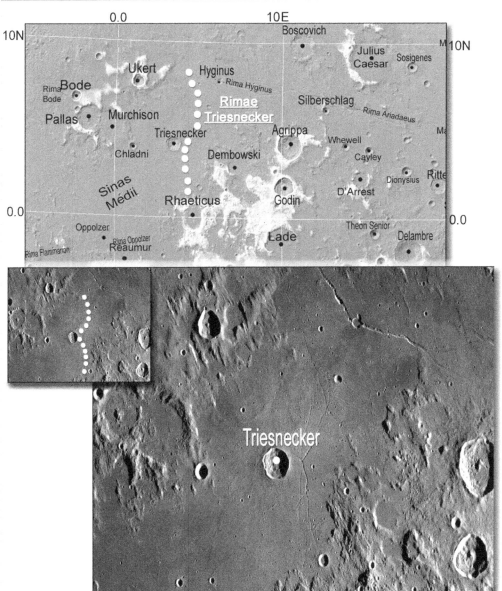

Rimae

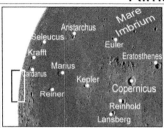

Rimae Vasco da Gama	
Lat 11.55N	Long 84.03W
Diameter	10.39Km
Notes:	

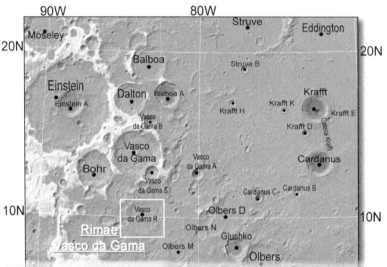

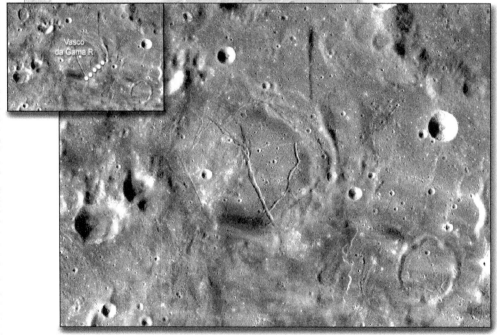

Rimae

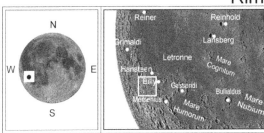

Rimae Zupus	
Lat 15.46S	Long 53.76W
Diameter	130.0Km
Notes:	

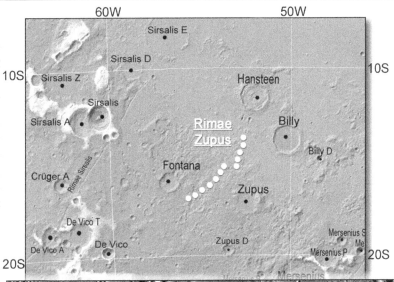

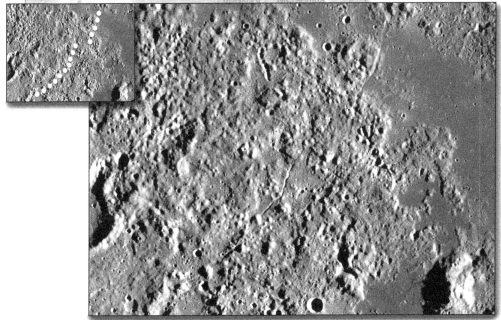

Rupes

Rupes

In the lunar context **_Rupes_** usually refers to fault-like features. The faults may occur in response to later tectonic events on the Moon, which then form through thrusting of one side of the surface over another. The most recognised faults on the Moon are Rupes Recta and Rupes Altai, however, some given below e.g. Rupes Kelvin or Rupes Mercator may not be so obvious. The word 'scarp' is sometimes used in relation to Rupes.

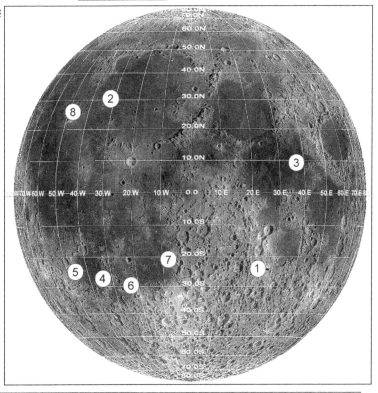

No.	Feature	No.	Feature	No.	Feature
1	Altai	4	Kelvin	7	Recta
2	Boris	5	Liebig	8	Toscanelli
3	Cauchy	6	Mercator		

Note: The above list of Rupes represents those given in the International Astronomical Union list, but there are plenty more on the Nearside's face that may in the future be given official designations.

Rupes

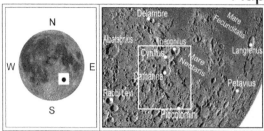

Rupes Altai	
Lat 24.32S	Long 23.12E
Diameter	545.19Km
Notes:	

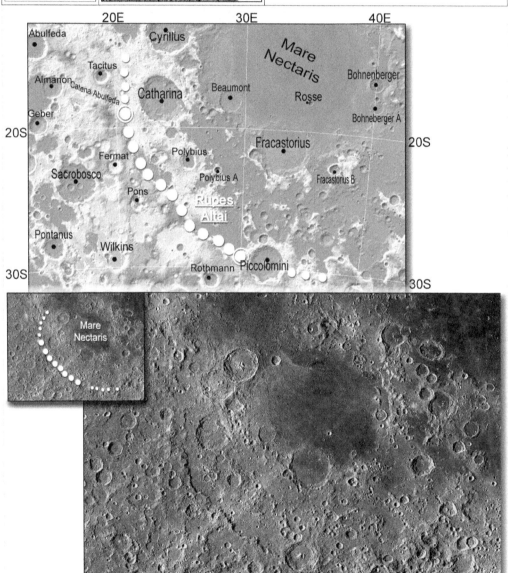

Rupes

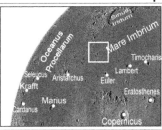

Rupes Boris	
Lat 30.67N	Long 33.6W
Diameter	8.58Km
Notes:	

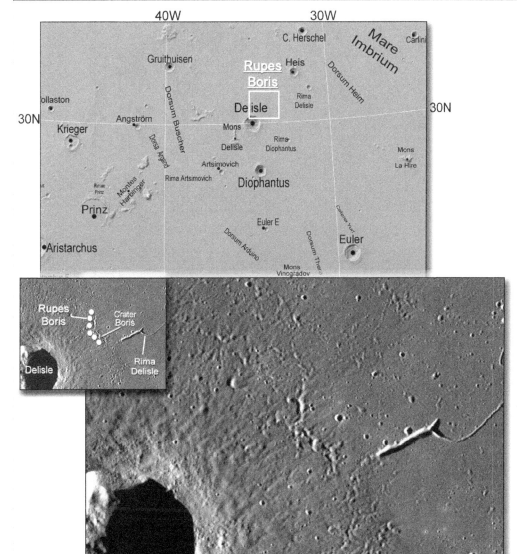

Rupes

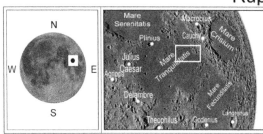

Rupes Cauchy	
Lat 9.31N	Long 37.08E
Diameter	169.85Km
Notes:	

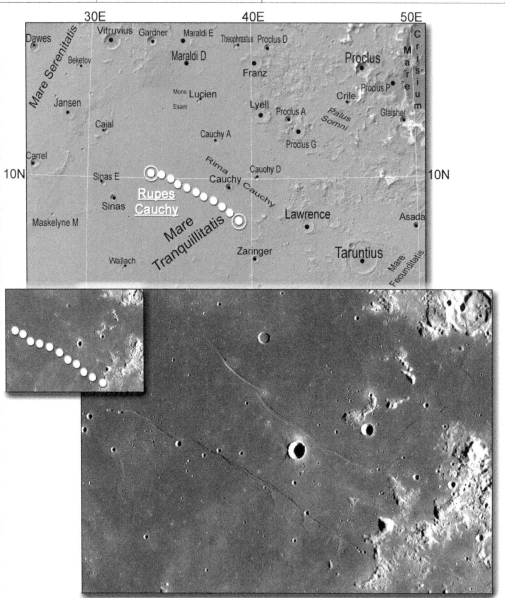

278

Rupes

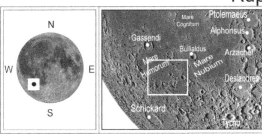

Rupes Kelvin	
Lat 28.03S	Long 33.17W
Diameter	85.92Km
Notes:	

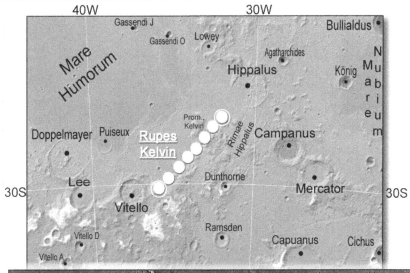

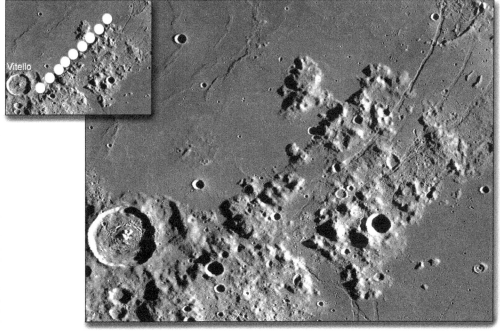

Rupes

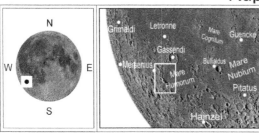

Rupes Liebig	
Lat 25.14S	Long 45.92W
Diameter	144.78Km
Notes:	

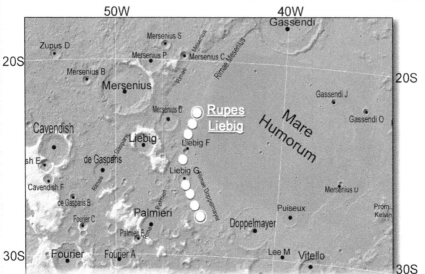

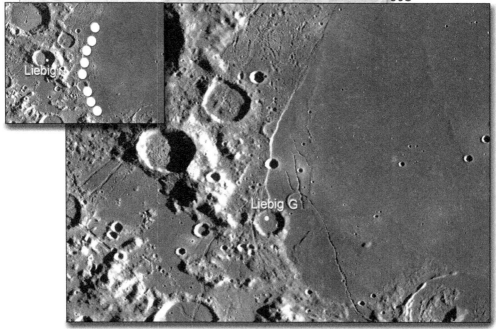

Rupes

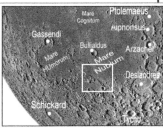

Rupes Mercator	
Lat 30.21S	Long 22.84W
Diameter	132.4Km
Notes:	

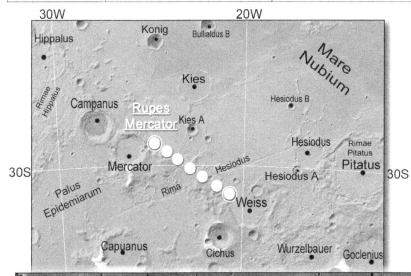

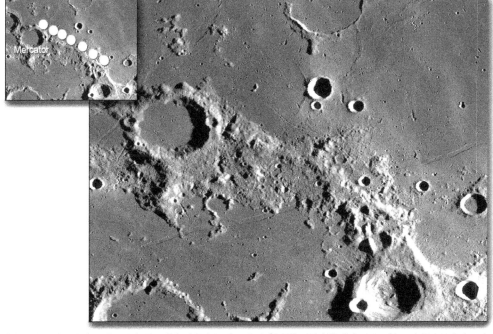

Rupes

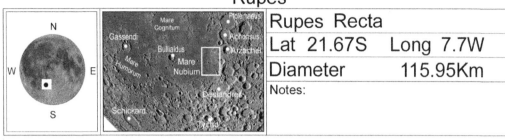

Rupes Recta	
Lat 21.67S	Long 7.7W
Diameter	115.95Km
Notes:	

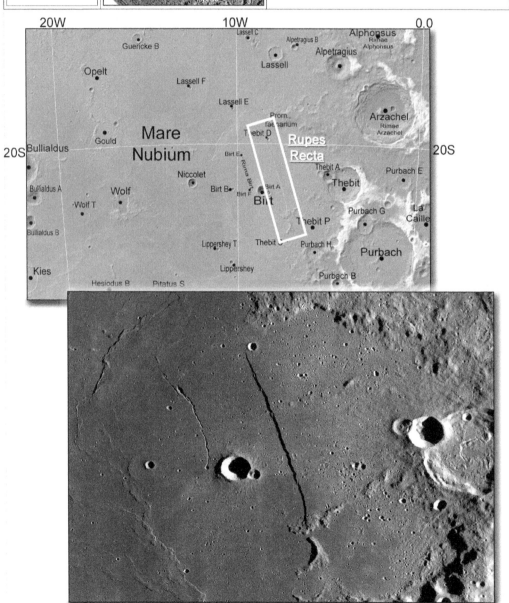

Rupes

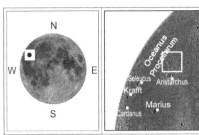

Rupes Toscanelli	
Lat 26.97N	Long 47.53W
Diameter	50.14Km
Notes:	

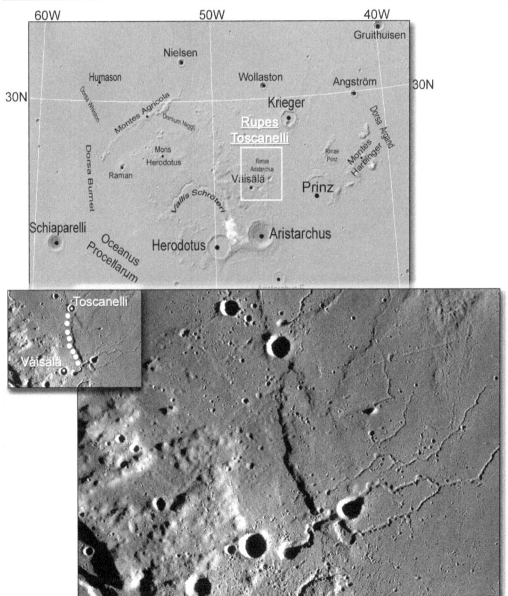

Sinus

Sinus

Translated from the Latin, *Sinus* stands for 'bay'. Bays on Earth are usually seen as large bodies of water connected to oceans and inlets, however, in context to the Moon they are seen as patches of lava connected to the Maria - the 'seas' of the Moon, and the mountain-like rims associated with them.

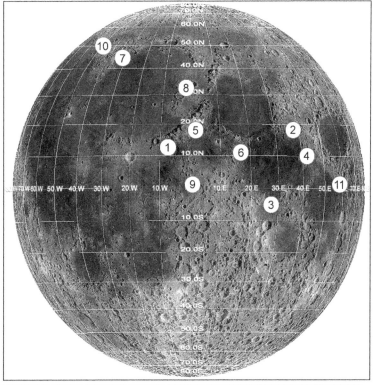

No.	Feature	No.	Feature	No.	Feature
1	Aestuum	5	Fidei	9	Medii
2	Amoris	6	Honoris	10	Roris
3	Asperitatis	7	Iridum	11	Successus
4	Concordiae	8	Lunicus		

Note: The above list of Sinus represents those given in the International Astronomical Union list, but there are plenty more on the Nearside's face that may in the future be given official designations.

Sinus

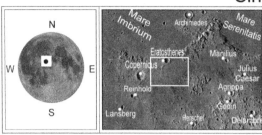

Sinus Aestuum	
Lat 12.1N	Long 8.34W
Diameter	316.5Km
Notes:	

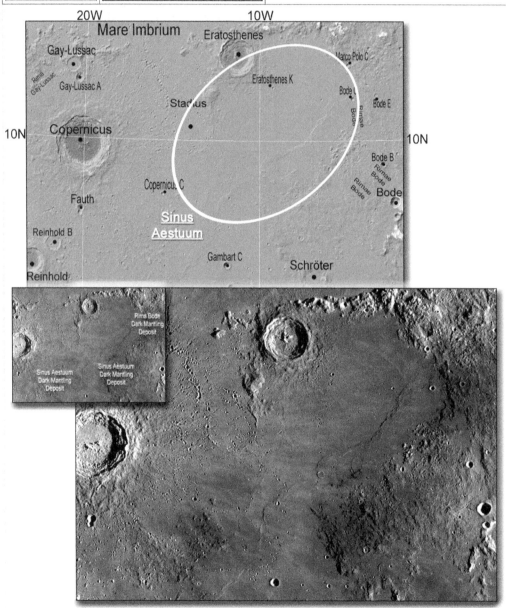

Sinus

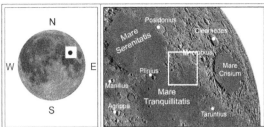

Sinus Amoris

Lat 19.92N	Long 37.29E
Diameter	189.1Km

Notes:

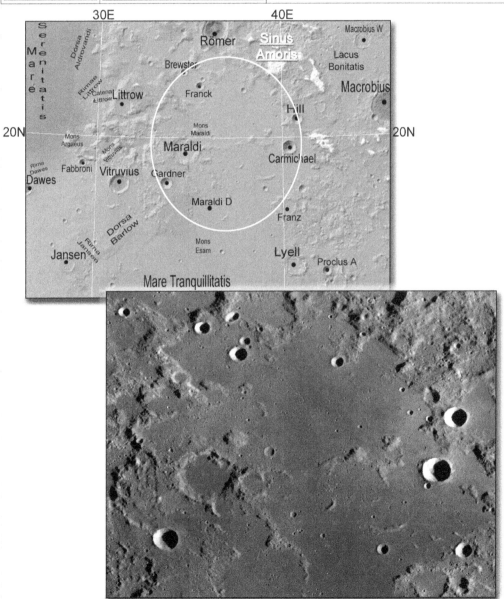

Sinus

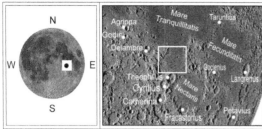

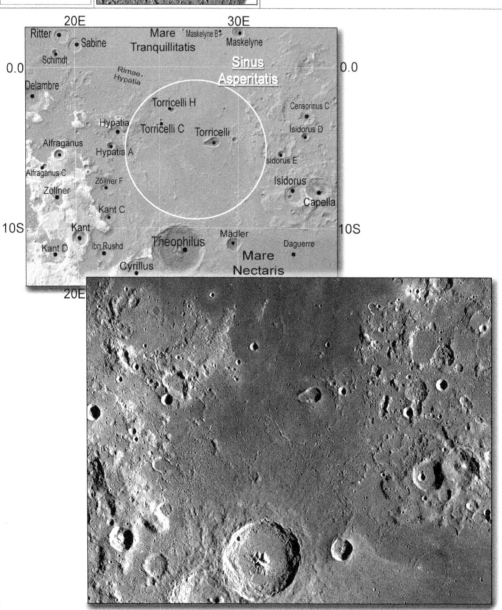

Sinus

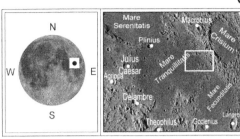

Sinus Concordiae	
Lat 10.98N	Long 42.47E
Diameter	159.03Km
Notes:	

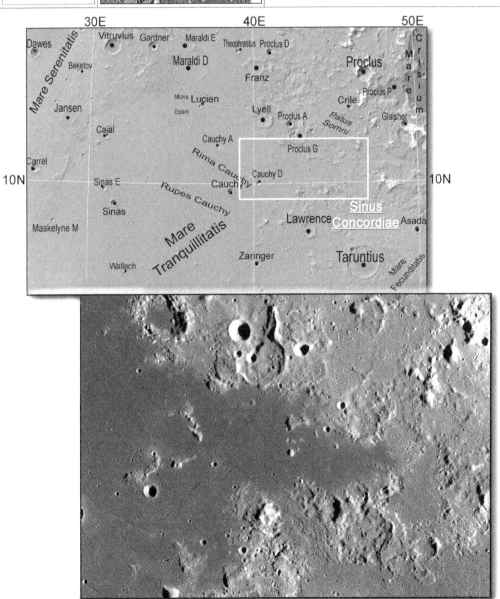

Sinus

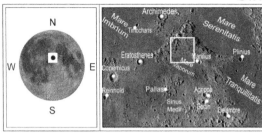

Sinus Fidei	
Lat 17.99N	Long 2.04E
Diameter	70.7Km
Notes:	

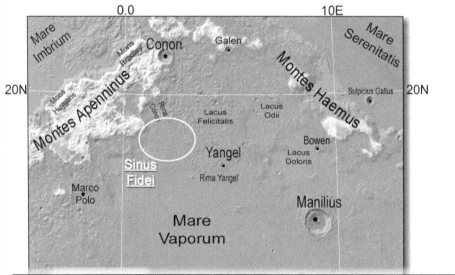

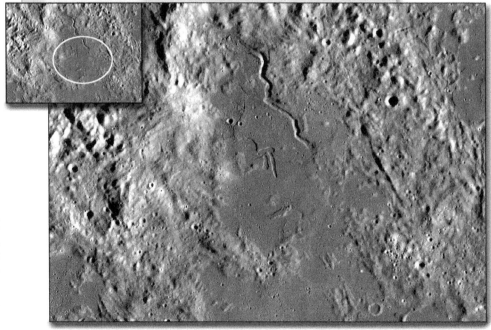

Sinus

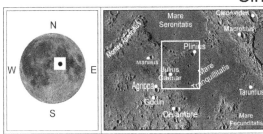

Sinus Honoris	
Lat 11.72N	Long 17.87E
Diameter	111.61Km
Notes:	

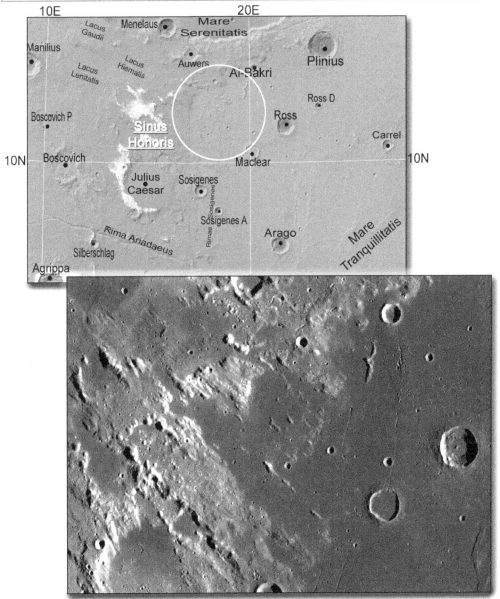

Sinus

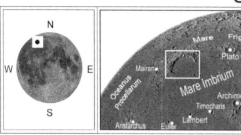

Sinus Iridum	
Lat 45.01N	Long 31.67W
Diameter	249.29Km
Notes:	

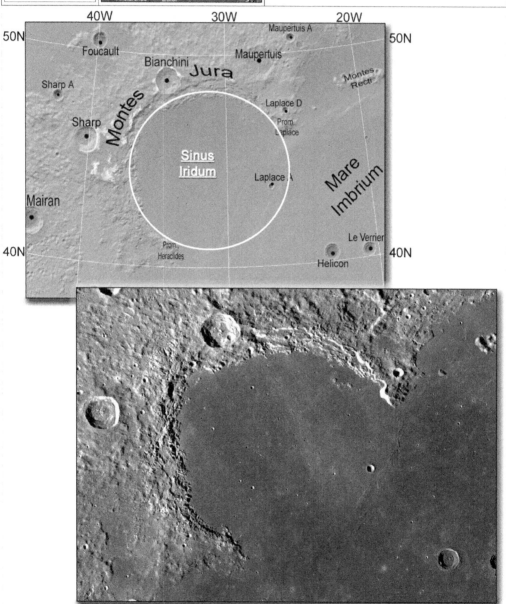

Sinus

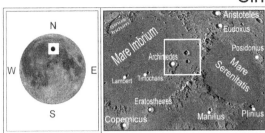

Sinus Lunicus	
Lat 32.36N	Long 1.85W
Diameter	119.18Km
Notes:	

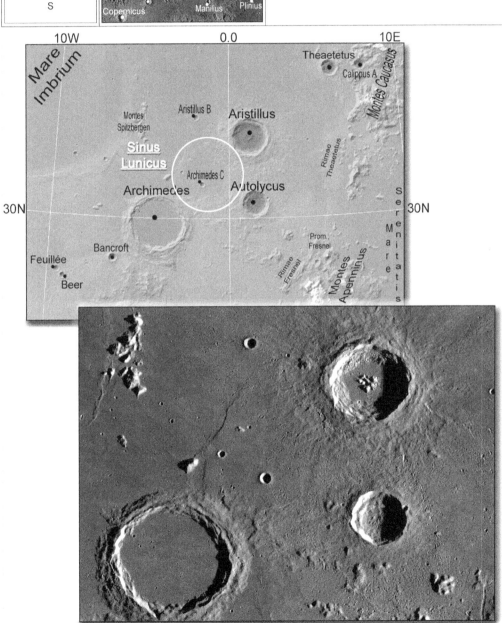

Sinus

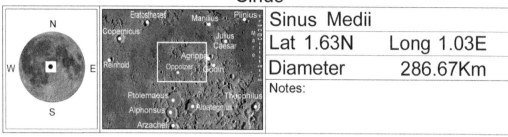

Sinus Medii	
Lat 1.63N	Long 1.03E
Diameter	286.67Km
Notes:	

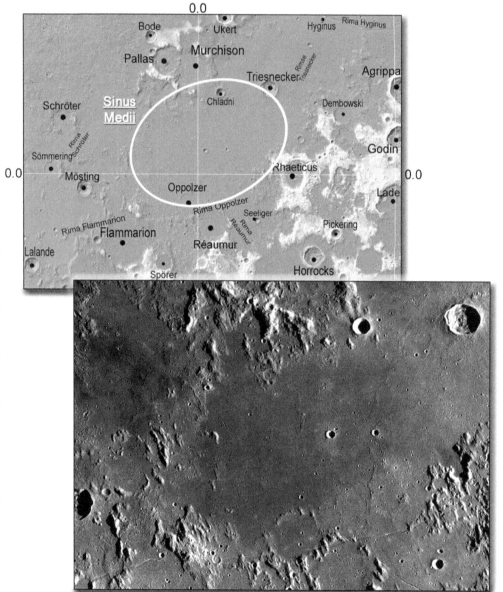

294

Sinus

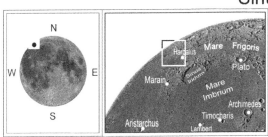

Sinus Roris	
Lat 50.26N	Long 50.86W
Diameter	195.04Km
Notes:	

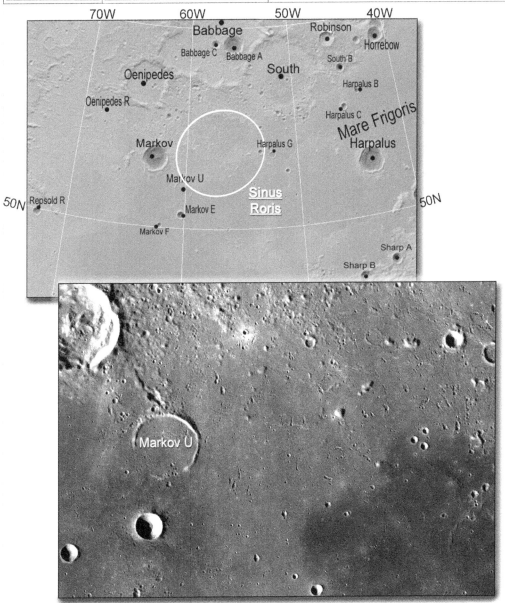

Sinus

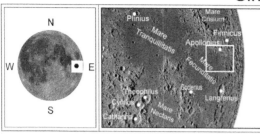

Sinus Successus	
Lat 1.12N	Long 58.52E
Diameter	126.65Km
Notes:	

Vallis

Vallis translates from the Latin into 'valley'. The only two Vallis that appear valley-like would be Vallis Altai and Vallis Schröteri, as all others are more 'chain-like' to 'gouged-out' features on the lunar surface.

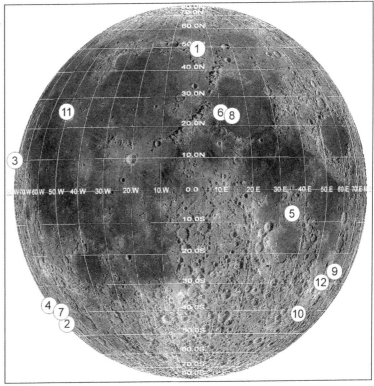

No.	Feature	No.	Feature	No.	Feature
1	Alpes	5	Capella	9	Palitzsch
2	Baade	6	Christel	10	Rheita
3	Bohr	7	Inghirami	11	Schröteri
4	Bouvard	8	Krishna	12	Snellius

Note: The above list of Vallis represents those given in the International Astronomical Union list, but there are plenty more on the Nearside's face that may in the future be given official designations.

Vallis

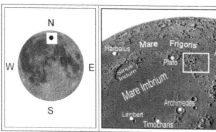

Vallis Alpes	
Lat 49.21N	Long 3.63E
Diameter	155.42Km
Notes:	

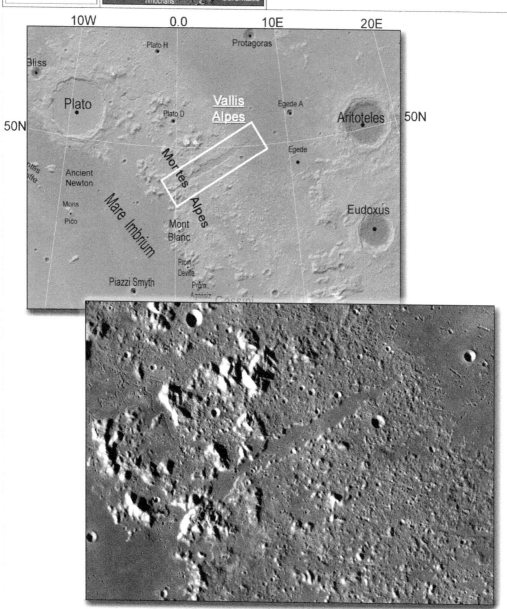

Vallis

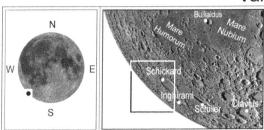

Vallis Baade	
Lat 45.55S	Long 77.23W
Diameter	206.79Km
Notes:	

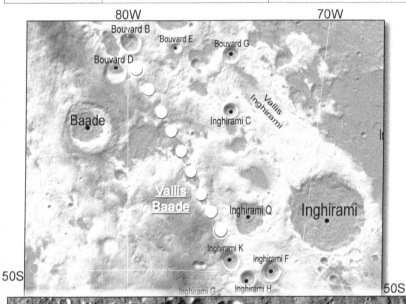

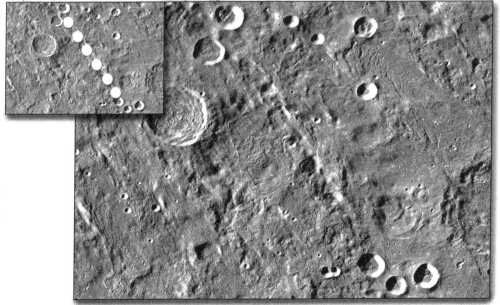

Vallis

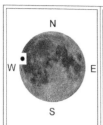

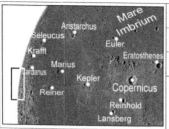

Vallis Bohr	
Lat 10.25N	Long 88.86W
Diameter	95.32Km
Notes:	

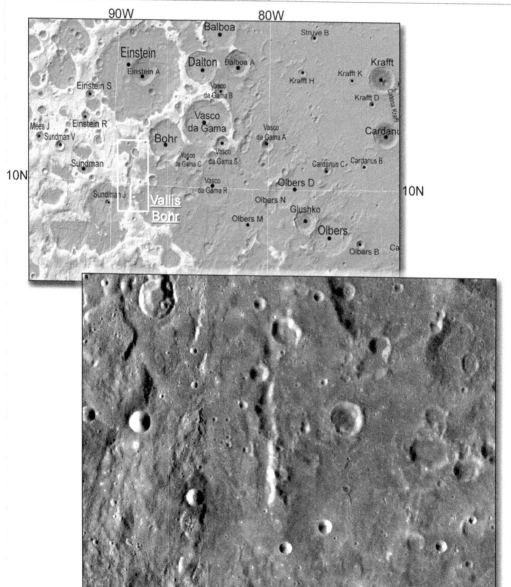

Vallis

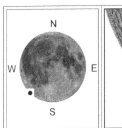

Vallis Bouvard	
Lat 38.45S	Long 82.32W
Diameter	287.92Km
Notes:	

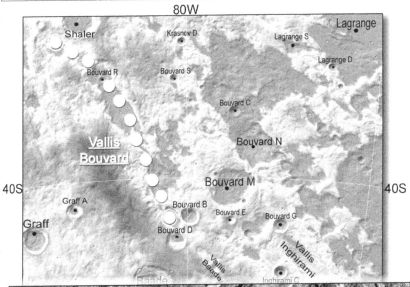

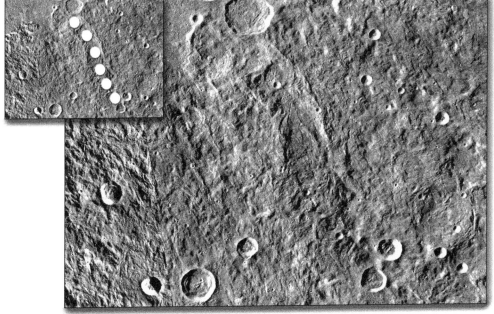

Vallis

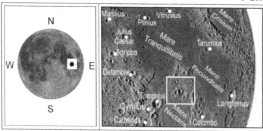

Vallis Capella	
Lat 7.39S	Long 35.04E
Diameter	106.28Km
Notes:	

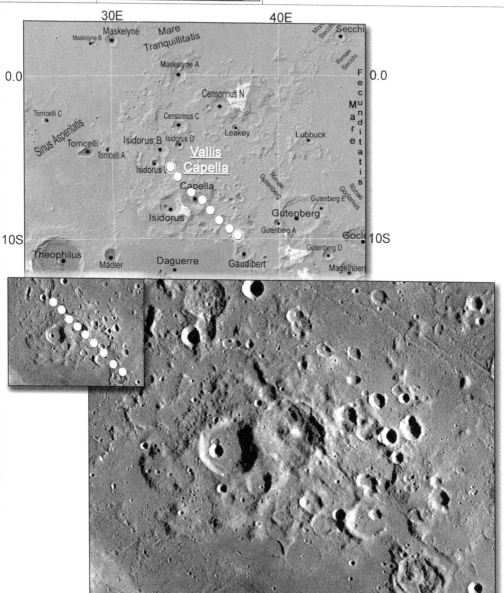

Vallis

Vallis Christel	
Lat 24.54N	Long 11.08E
Diameter	2.1Km
Notes:	

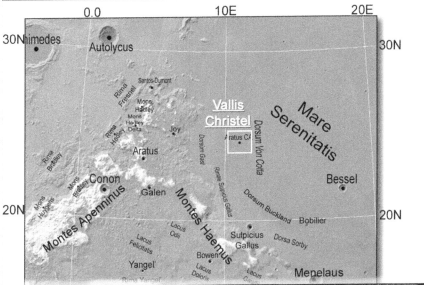

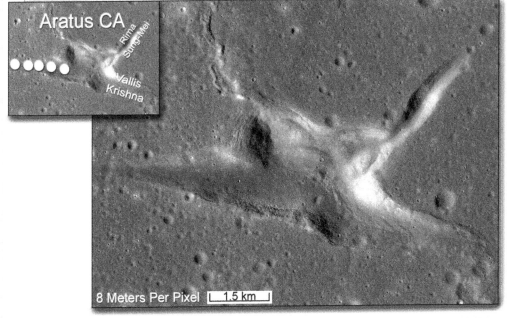

Aratus CA

Rima Sung-Mei

Vallis Krishna

8 Meters Per Pixel   1.5 km

Vallis

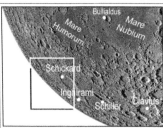

Vallis Inghirami	
Lat 43.95S	Long 72.59W
Diameter	145.08Km
Notes:	

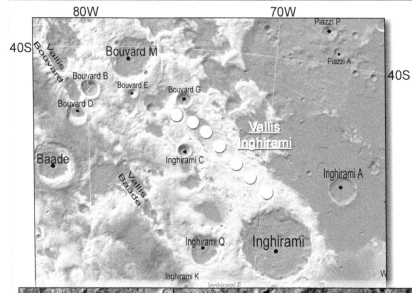

Vallis

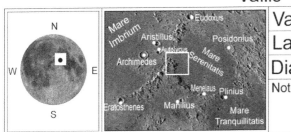

Vallis Krishna	
Lat 24.54N	Long 11.26E
Diameter	2.9Km
Notes:	

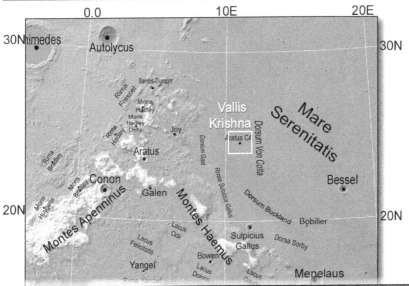

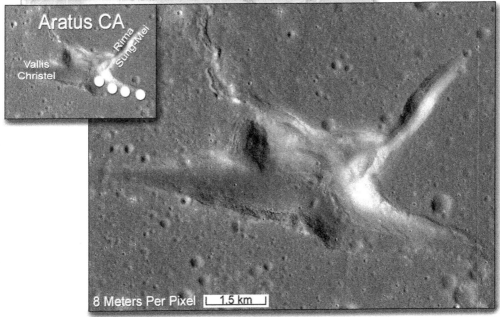

Aratus CA

Rima Sung-Mei

Vallis Christel

8 Meters Per Pixel   1.5 km

Vallis

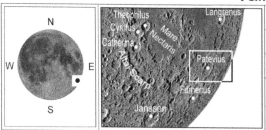

Vallis Palitzsch	
Lat 26.16S	Long 64.64E
Diameter	110.5Km
Notes:	

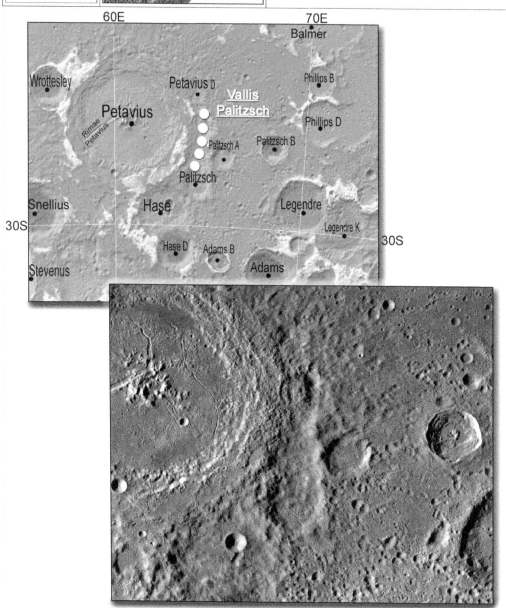

306

Vallis

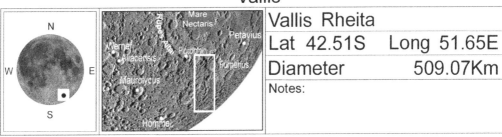

		Vallis Rheita
		Lat 42.51S Long 51.65E
		Diameter 509.07Km
		Notes:

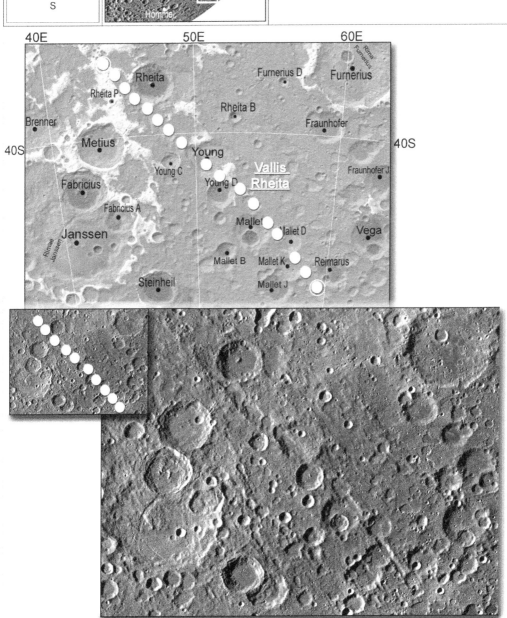

Vallis

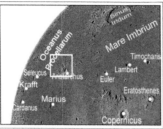

Vallis Schröteri	
Lat 26.16N	Long 51.58W
Diameter	185.32Km
Notes:	

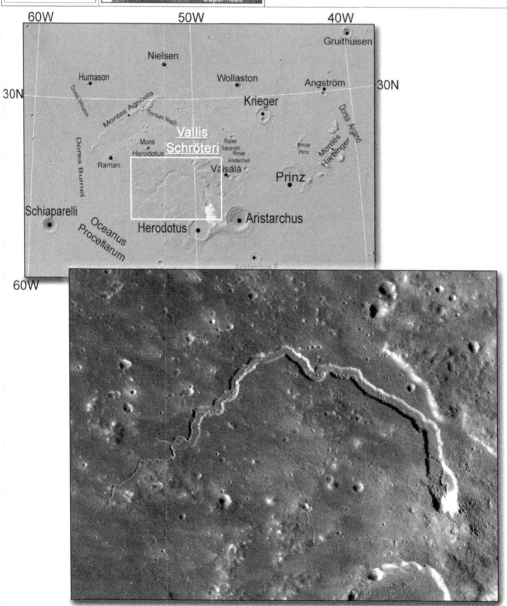

Vallis

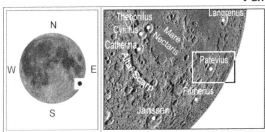

Vallis Snellius

Lat 30.93S	Long 57.84E
Diameter	640.0Km

Notes:

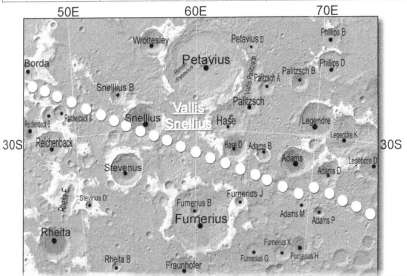

Theophilus
Cyrilus
Catherina
Mare
Nectaris
Altai Scarp
Langrenus
Petavius
Furnerius
Janssen

50E   60E   70E

Wrottesley
Petavius D
Phillips B
Borda
Petavius
Rimae Petavius
Vallis Palitzsch
Palitzsch A
Palitzsch B
Phillips D
Snellius B
Vallis Snellius
Palitzsch
Reichenbach B
Reichenbach A
Snellius
Hase
Legendre
Legendre K
30S   Reichenbach   Hase D   Adams B   30S
Stevenus
Adams
Legendre D
Rheita E
Stevinus D
Furnerius J
Adams D
Stevinus D
Furnerius B
Adams M
Adams P
Furnerius
Rheita
Furnerius K
Rheita B
Fraunhofer
Furnerius G
Furnerius H

309

Concentric Craters

Concentric Craters

The formation of ***Concentric craters*** is still misunderstood today. Are they showing one small impact crater within another, slightly larger impact crater? Are they produced by a single impactor striking layered target rock (affecting rebound properties)? Are they due to simultaneous impacts, volcanic extrusion or igneous intrusion. The list of these objects is growing each year - the few given here to serve as an introduction.

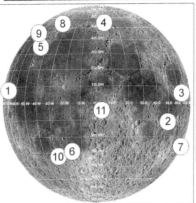

No.	Region
1	Cavalerius
2	Crozier
3	Dubyago
4	Egede, Aristoteles
5	Gruithuisen, Mairan
6	Hesiodus, Nubium
7	Humboldt
8	J. Herschel, Frigoris
9	Louville, Sharp
10	Marth, Capuanus
11	Réaumus, Gyldén

Cavalerius region

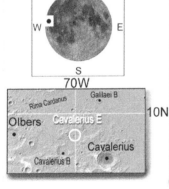

70W

10N

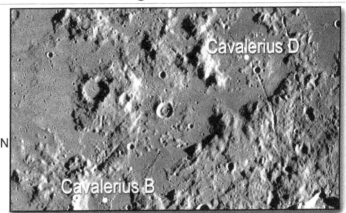

Crozier region

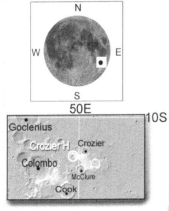

50E

10S

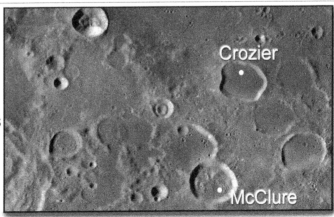

Concentric Craters

Dubyago region

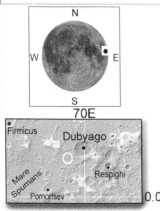

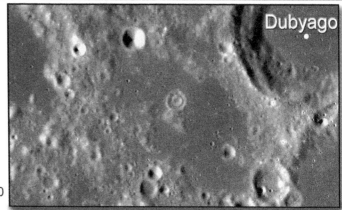

70E

Firmicus
Dubyago
Mare Spumans
Respighi
Pomortsev

0.0

Dubyago

Egede, Aristoteles region

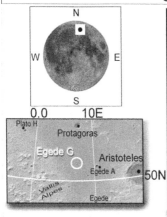

0.0   10E

Plato H
Protagoras
Egede G
Aristoteles
Egede A
Vallis Alpes
Egede

50N

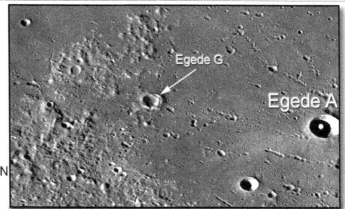

Egede G

Egede A

Gruithuisen, Mairan region

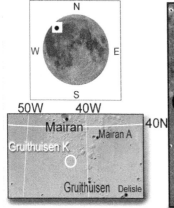

50W   40W

Mairan
Mairan A
Gruithuisen K
Gruithuisen   Delisle

40N

Concentric Craters

Hesiodus, Nubium Basin region

20W 10W

Mare
Nubium

Pitatus

Rima Hesiodus

Weiss

Hesiodus A

Gauricus

Wurzelbaur

30S

Humboldt region

80E 90E

Hecataeus

Curie

Humboldt

Phillips

Barnard

30S

J. Herschel, Frigoris region

50W 40W 30W

J. Herschel

J.Herschel F

Mare
Frigoris

60N

Harpalus

LaCondamine

Concentric Craters

Louville, Sharp region

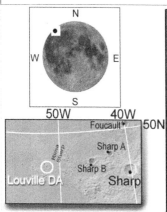

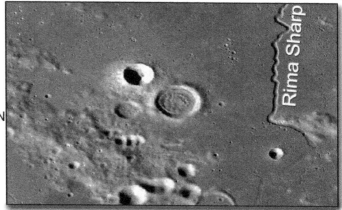

Marth, Capuanus region

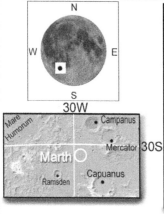

Réaumur, Gyldén region

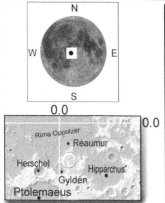

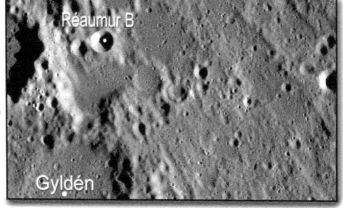

Dark Halo Craters

Dark Halo Craters

Dark Halo Craters display dark-like deposits around their exterior. In some cases the deposits appear to be very dark (see Alphonsus region) and dispersed oddly around the central crater (or possible related vent), while in others (see Theophilus region) the deposits are distributed equally around their craters. Each display may say something about their original formation (volcanic or impact), and each about sub-surface unit plains.

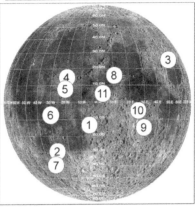

No.	Region
1	Alphonsus
2	Capuanus, Mercator
3	Cleomedes
4	Copernicus (north)
5	Copernicus (south)
6	Euclides
7	Hainzel
8	Manilius
9	Theophilus
10	Torricelli
11	Triesnecker

Alphonsus region

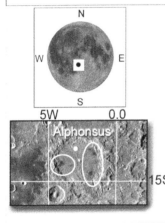

Capuanus, Mercator region

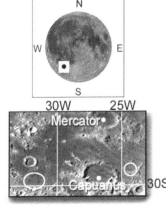

Dark Halo Craters

Cleomedes region

Copernicus (north) region

Copernicus (south) region

Dark Halo Craters

Euclides region

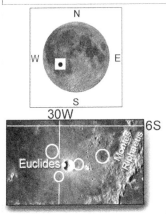

Hainzel region

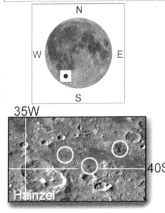

Manilius region

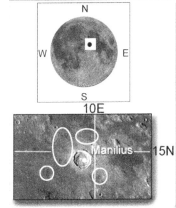

Dark Halo Craters

Theophilus region

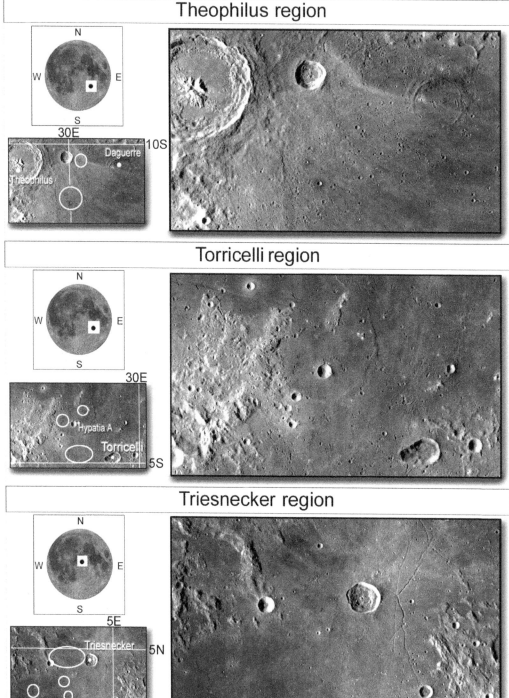

N
W E
S

30E

10S

Daguerre

Theophilus

Torricelli region

N
W E
S

30E

Hypatia A

Torricelli

5S

Triesnecker region

N
W E
S

5E

Triesnecker

5N

Domes

Domes

Appearing like 'dimples' on the lunar surface, ***Domes*** represent the volcanic aspect similar as to how we view volcanoes on Earth today. Due to several factors e.g. lava viscosity or composition, rate and duration in effusion of said, they can form like shield-like volcanoes with somewhat steep slopes and small diameters, or shallower slopes with larger diameters as seen in the intrusive types like the Valentine dome.

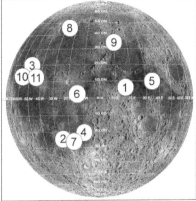

No.	Region
1	Arago, Tranquillitatis
2	Archytas, Frigoris
3	Aristarchus
4	Birt, Rupes Recta
5	Cauchy, Tranquillitatis
6	Gambart
7	Kies, Nubium
8	Laplace, Sinus Iridum
9	Linné, Serenitatis
10	Marius Hills
11	Mons Rümker

Arago, Tranquillitatis region

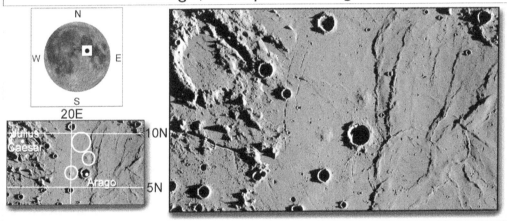

Archytas, Frigoris region

Domes

Aristarchus region

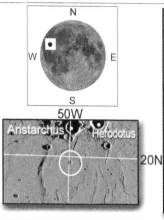

50W

Aristarchus Herodotus

20N

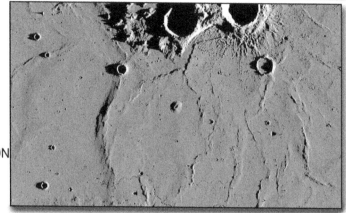

Birt, Rupes Recta region

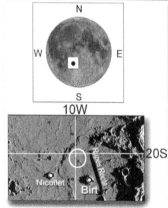

10W

Nicollet Birt Rupes Recta

20S

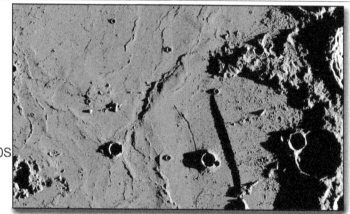

Cauchy, Tranquillitatis region

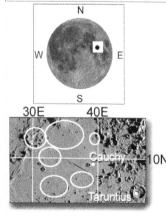

30E 40E

Cauchy

Taruntius

10N

Domes

Gambart region

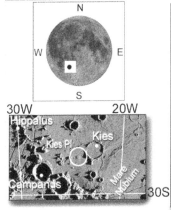

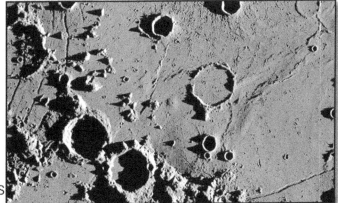

Kies, Nubium region

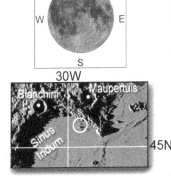

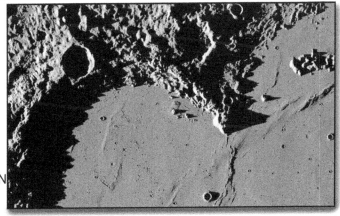

Laplace, Sinus Iridum region

Domes

Linné, Serenitatis region

Map labels: N, W, E, S; 5E, 15E; Linné H, Linné F, Valentine, Linné B, Linné A; 30N

Marius Hills region

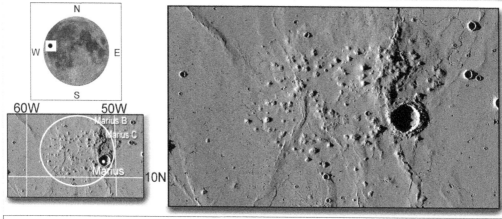

Map labels: N, W, E, S; 60W, 50W; Marius B, Marius C, Marius; 10N

Mons Rümker region

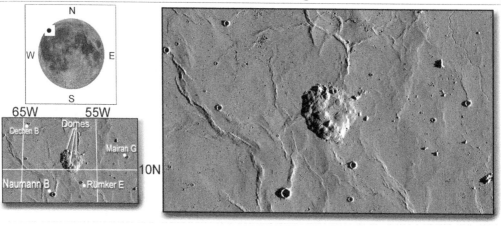

Map labels: N, W, E, S; 65W, 55W; Dechen B, Domes, Mairan G, Naumann B, Rümker E; 10N

Floor Fractured Craters

Floor Fractured Craters

Floor Fractured Craters are believed to form when shallow magmatic activities under extreme pressure cause their floors to uplift - leading to radial, concentric or polygonal type fractures. Proximity to local mares or highlands may also play a role in the fracturing process as crustal thickness affects pressure and uplift. A large number of FFCs may be pre-mare impact craters whose floors were modified by mare flooding epochs.

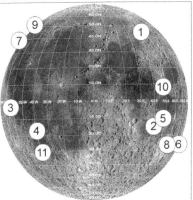

No.	Region
1	Atlas
2	Bohnenberger
3	Damoiseau
4	Gassendi
5	Goclenius
6	Humboldt
7	Lavoisier, Bunsen
8	Petavius
9	Repsold
10	Taruntius
11	Vitello

Atlas region

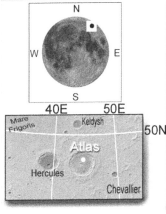

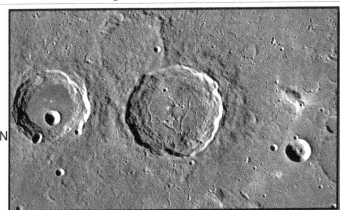

Bohnenberger region

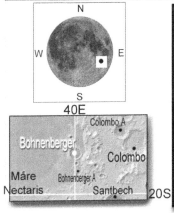

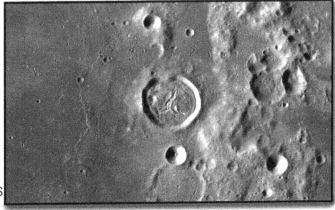

Floor Fractured Craters

Damoiseau region

N
W E
S
60W

Hermann
Damoiseau
Grimaldi
Damoiseau A
Sirsalis E
0.0

Gassendi region

N
W E
S
40W

Letronne
Herigonius
Gassendi B
Gassendi A
Gassendi
Mersenius
Mare Humorum
20S

Goclenius region

N
W E
S
40E 50E

Mare Fecunditatis
Gutenberg
Goclenius
Mare Nectaris
Magelhaens
Colombo
10S

324

Floor Fractured Craters
Humboldt region

Lavoisier, Bunsen region

Petavius region

Floor Fractured Craters

Repsold region

Taruntius region

Vitello region

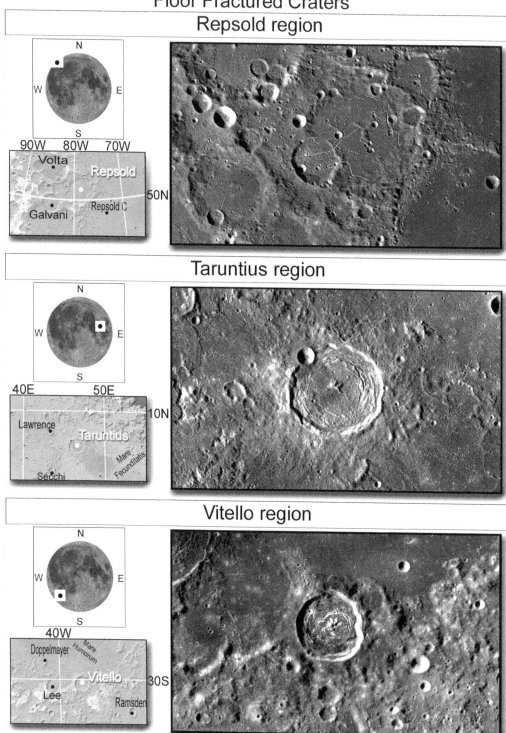

Ghost Craters

Ghost Craters

The ***Ghost Craters*** featured below have one thing in common - they all lie within mares. Take away the mare deposits and one would see a normal-looking crater underneath. GCs thus represent craters whose floors and rims were emplaced by lava deposits, leaving just their rims exposed to signify they once existed. Their local presence may have affected distribution of said lavas, and also later effects with surface wrinking.

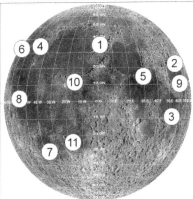

No.	Region
1	Aristillus, Imbrium
2	Eimmart, Crisium
3	Goclenius, Fecunditatis
4	Gruithuisen, Mairan
5	Jansen, Tranquillitatis
6	Lichtenberg, Oceanus Proc.,
7	Liebig, Humorum
8	Reiner, Reiner Gamma
9	Shapely, Crisium
10	Stadius, Sinus Aestuum
11	Wolf, Nubium

Aristillus, Imbrium region

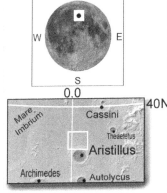

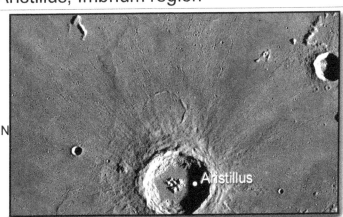

Eimmart, Crisium region

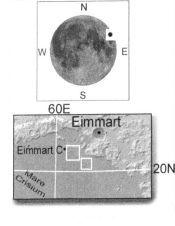

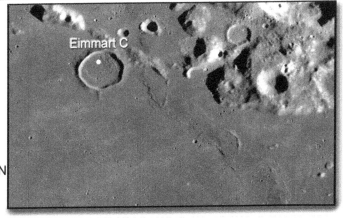

Ghost Craters

Goclenius, Fecunditatis region

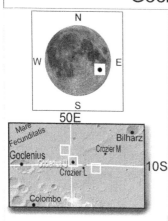

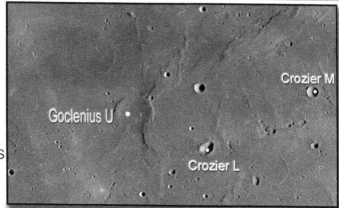

Crozier M

Goclenius U

Crozier L

Gruithuisen, Mairan region

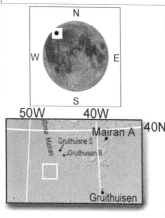

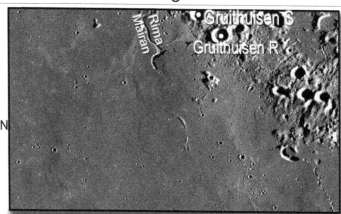

Rima Mairan

Gruithuisen S

Gruithuisen R

Jansen, Tranquillitatis region

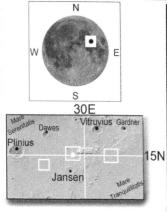

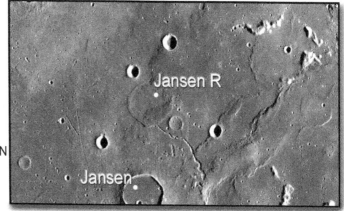

Jansen R

Jansen

Ghost Craters

Lichtenberg, Oceanus Procellarum region

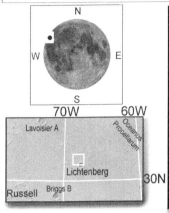

Lichtenberg

Liebig, Humorum region

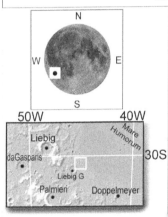

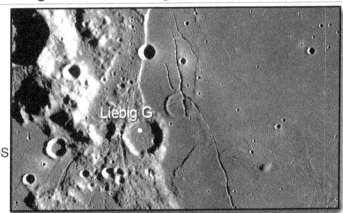

Liebig G

Reiner, Reiner Gamma region

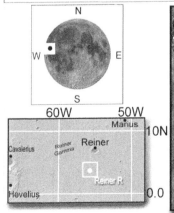

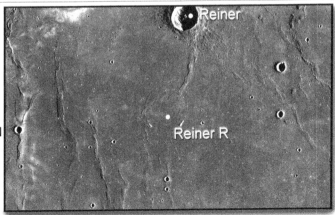

Reiner

Reiner R

Ghost Craters

Shapley, Crisium region

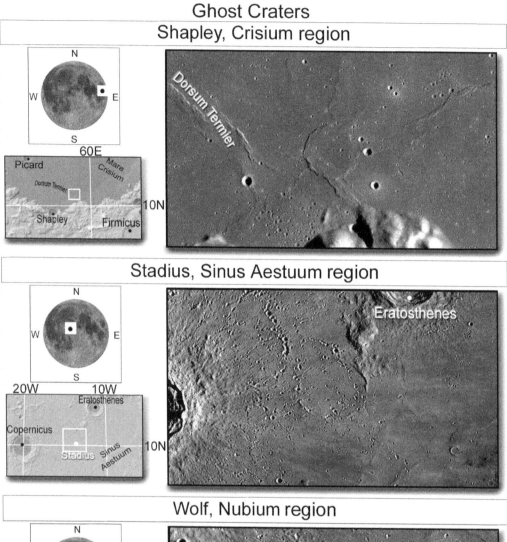

Stadius, Sinus Aestuum region

Wolf, Nubium region

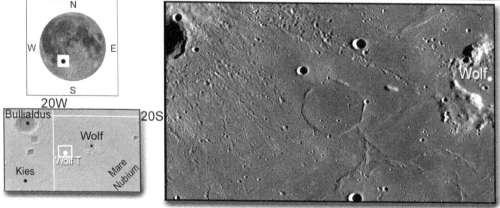

Miscellaneous

Miscellaneous

The ***Miscellaneous*** list given below represent a general introduction to features not so obvious on the lunar surface. Some will 'pop out' during an observing run, while others may require suitable lighting conditions, if not an optical instrument using high magnification - the 'Ina Structure' being the most challenging. There are more features to be seen - all that is required however is a keen eye and interest.

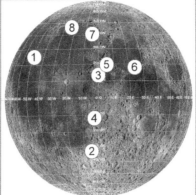

No.	Region
1	Aristarchus Plateau
2	Cassini's Bright Spot
3	Dark Deposits
4	Imbrium Sculpture
5	Ina Structure
6	Lava Divides
7	Piazzi Smyth V
8	Rille - Hidden and Exposed

Aristarchus Plateau

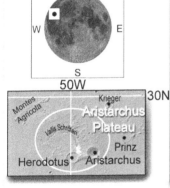

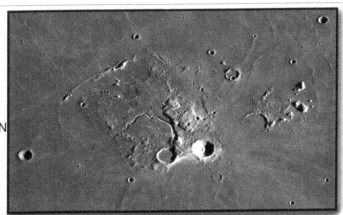

Cassini's Bright Spot

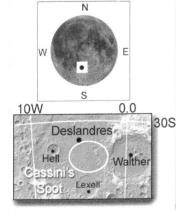

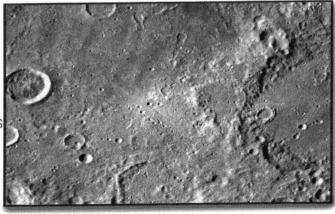

Miscellaneous
Dark Deposits

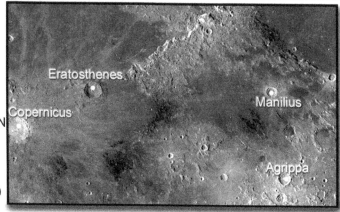

Imbrium Sculpture

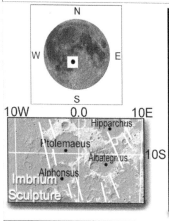

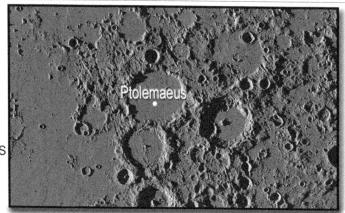

Ina Structure

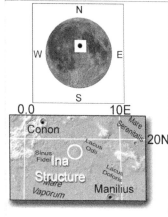

Miscellaneous

Lava Divides

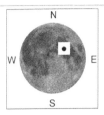

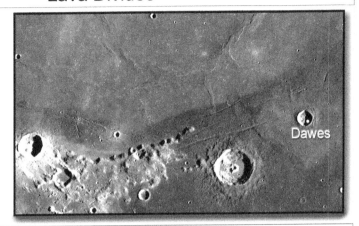

Piazzi Smyth V

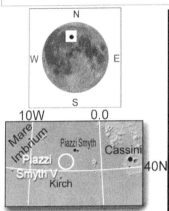

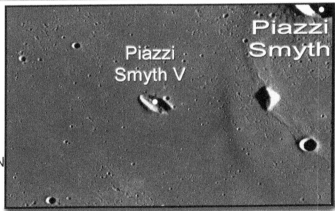

Rille - Hidden and Exposed

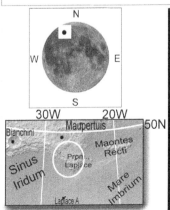

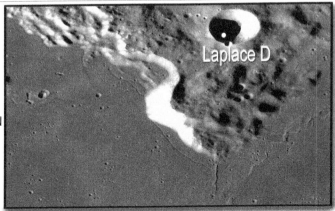

333

Craters With Rays

Craters With Rays

Craters With Rays are signatures of some of the youngest craters on the Moon. They are catagorised into two main types: firstly, those rays having a more 'immature' aspect - where fresh, sub-surface material has been brought to the surface through impact; secondly, those rays where 'maturity' in composition (e.g. highland's crust) whose brightness may contrast against a more darker surface, such as a mare. Rayed craters can occur through the primary impactor, or from secondaries from said that usually produce a smaller ray system around them. The system of rays usually can last for less than a billion years, which slowly degrade in albedo into their target terrain. The list below serves as an introduction to these wonderful features, but there are plenty more to be found.

No.	Anaxagoras	No.	Byrgius A	No.	Copernicus
1	73.48N, 10.17W	2	24.55S, 63.81W	3	9.62N, 20.08W

No.	Glushko	No.	Kepler	No.	Langrenus
4	8.11N, 77.67W	5	8.12N, 38.01W	6	8.86S, 61.04E

Craters With Rays

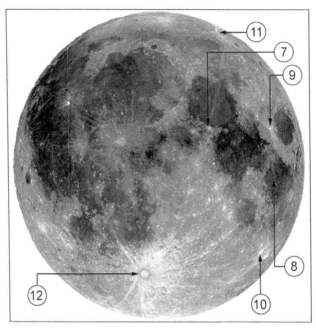

No.	Menelaus
7	16.26N, 15.93E

No.	Messier/Messier A
8	1.9S, 47.65E

No.	Proclus
9	16.09N, 46.89E

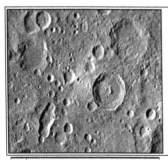

No.	Stevinus A
10	31.86S, 51.65E

No.	Thales
11	61.74N, 50.27E

No.	Tycho
12	43.3S, 11.22W

Simultaneous Impact Craters

Simultaneous Impact Craters

Simultaneous Impact Craters as the title suggests are craters that formed on impact on the lunar surface, roughly, at the same time. Identification of such features is usually seen as a straight, septum-like ridge 'connecting' the craters involved, which in most cases presented below appear relatively sharp. Variations, however, can occur too, where a pinch-like feature between the two craters turns up e.g. Plato K and Plato KA.

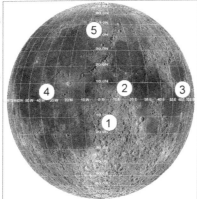

No.	Region
1	Argelander
2	Ariadaeus
3	Daly
4	Encke
5	Plato

Argelander region

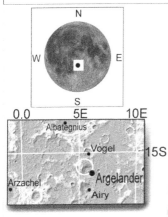

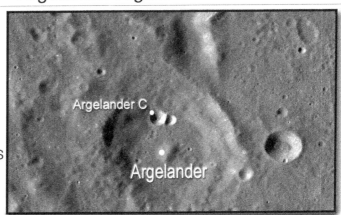

Ariadaeus region

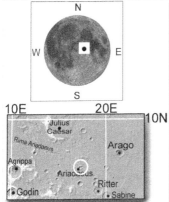

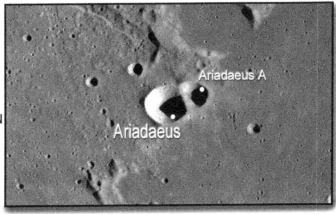

Simultaneous Impact Craters

Daly region

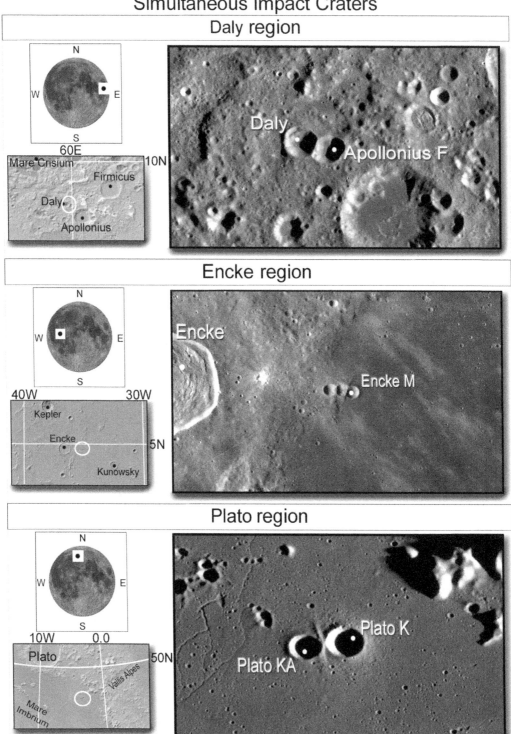

Encke region

Plato region

Swirls

Swirls

Inappropriately named, **_Swirls_** really don't appear 'swirl-like' at all, but more like variations in albedo difference on some parts of the lunar surface. Their formation still isn't quite understood, however, amongst the most popular of theories is one where concentrated magnetic fields in their locality may be deflecting ions from the solar flux from darkening the lunar soil (regolight).

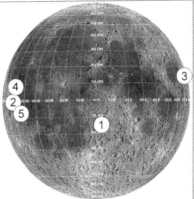

No.	Region
1	Airy, Parrot
2	Lohrmann
3	Marginis
4	Reiner
5	Sirsalis

Airy, Parrot region

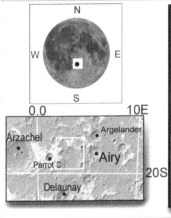

Lohrmann region

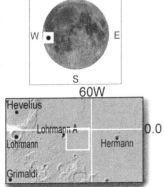

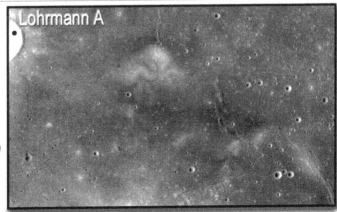

Swirls

Marginis region

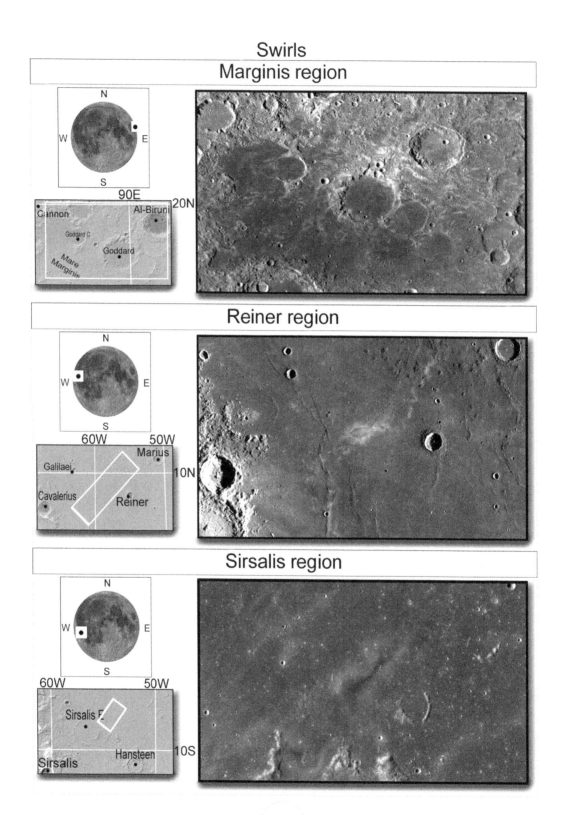

N
W   E
S
90E

Cannon   Al-Biruni
20N
Goddard C
Goddard
Mare
Marginis

Reiner region

N
W   E
S
60W   50W

Marius
Galilaei
10N
Cavalerius   Reiner

Sirsalis region

N
W   E
S
60W   50W

Sirsalis E
Hansteen   10S
Sirsalis

Index

Index

Index

Index

Feature	Page

Index

Index

Index

Index

347

Index

Index

Index

Index

Index

Index

Index

Made in the USA
Las Vegas, NV
31 August 2021